COAL FLY ASH BENEFICIATION - TREATMENT OF ACID MINE DRAINAGE WITH COAL FLY ASH

Edited by **Segun A. Akinyemi**
and **Mugera W. Gitari**

Coal Fly Ash Beneficiation - Treatment of Acid Mine Drainage with Coal Fly Ash
http://dx.doi.org/10.5772/65147
Edited by Segun A. Akinyemi and Mugera W. Gitari

Contributors

Madhumita Roy, Roopali Roychowdhury, Pritam Mukherjee, Grzegorz Ludwik Golewski, Mugera Gitari, Segun Akinyemi, Leslie Petrik, Anita Etale, Nikita Tavengwa, Vusumzi Pakade, Ajit Behera, Soumya Mohapatra, Yıldırım İsmail Tosun, Deniz Sanliyuksel Yucel, Burcu Ileri, Sri Hartuti, Shinji Kambara, Akihiro Takeyama, Farrah Fadhillah Hanum, Erda Rahmilaila Desfitri

Published by InTech
Janeza Trdine 9, 51000 Rijeka, Croatia

Publishing Process Manager Maja Bozicevic
Technical Editor SPi Global
Cover InTech Design team

First published January, 2018
Printed in Croatia
Legal deposit, Croatia: National and University Library in Zagreb

Additional hard copies can be obtained from orders@intechopen.com

Coal Fly Ash Beneficiation - Treatment of Acid Mine Drainage with Coal Fly Ash, Edited by Segun A. Akinyemi and Mugera W. Gitari
p. cm.
Print ISBN 978-953-51-3752-8
Online ISBN 978-953-51-3753-5

For EU product safety concerns:
IN TECH d.o.o., Prolaz Marije Krucifikse Kozulić 3, 51000 Rijeka, Croatia,
info@intechopen.com or visit our website at intechopen.com.

Meet the editors

Editor, Dr. Segun A. Akinyemi had his undergraduate training in the Department of Geology, Ondo State University, now Ekiti State University, where he finished with a strong Second Class Upper division, in honours Geology (1995). He obtained both his MSc (2008) and PhD (2011) degrees in a reputable university, University of the Western Cape, South Africa. He won a German DAAD Postdoctoral Research Fellowship in Sub-Saharan Africa (2015–2016), which he undertook at the School of Environmental Science in the University of Venda, South Africa. Dr. Akinyemi has over 15 years of experience as a professional academic and research scientist with proven record of accomplishment in leading multidisciplinary research and development. He has conducted research in many areas of fossil fuel and environmental geochemistry, besides many other areas of applied geochemistry. He has also taught at both the undergraduate and graduate levels in Applied Geology since 2009.

Co-Editor, Prof. Mugera W. Gitari received his Bachelor of Education (Science) degree from Kenyatta University, Nairobi, Kenya, in 1994; Master of Science in Pure and Applied Chemistry from the University of Nairobi, Nairobi, Kenya, in 2000; and doctoral degree in Chemistry in 2006 from the University of the Western Cape, Cape Town, South Africa. Prof. Gitari has worked in tertiary education institutions for over 14 years and worked as a Postdoctoral Research Associate in the University of the Western Cape from 2006 to 2009. He is a full professor of Environmental/Analytical Chemistry at the School of Environmental Sciences, University of Venda, South Africa. His research interests are environmental remediation through beneficial and innovative utilization of industrial and mining waste products, development of functional materials for household water treatment and water quality monitoring and management. He has published 75 articles in peer-refereed journals, 5 book chapters and holds 4 patents in acid mine drainage and drinking water treatment. He has also mentored 21 MSc, PhD and postdoctoral candidates.

Contents

Preface

The book *Coal Fly Ash Beneficiation - Treatment of Acid Mine Drainage with Coal Fly Ash* deals with some of the significant successes achieved in treatment of acid mine drainage with coal fly. We hope the various topics in the book will enable our readers to better appreciate just some of the numerous scientific advancements on the subject matter. The topics being covered here are as follows: (1) "Introductory Chapter: Coal Fly Ash and Its Applications for Remediation of Acid Mine Drainage"; (2) "Metal Adsorption by Coal Fly Ash: The Role of Nano-sized Materials"; (3) Chemical Stabilization of Coal Fly Ash for Simultaneous Suppressing of As, B and Se Leaching"; (4) Evaluation of Ultrasound-assisted Modified Fly Ash for Treatment of Acid Mine Drainage"; (5) Treatment of Acid Mine Drainage with Coal Fly Ash: Exploring the Solution Chemistry and Product Water Quality"; (6) Phytoreclamation of Abandoned Acid Mine Drainage Site After Treatment with Fly Ash"; (7) Challenges in Recovery of Valuable and Hazardous Elements from Bulk Fly Ash and Options for Increasing Fly Ash Utilization"; (8) Thickener Water Neutralization by Mid-Bottom and Fly Ash of Thermal Power Plants and CO_2: Organic Humate Mud of AMD Treatment for Remediation of Agricultural Fields"; and (9) Fracture Toughness of Concrete Containing Fly Ash. These carefully selected topics are up-to-date, thoroughly referenced and literature cited to the present time and indeed review the relevant topics in each of these subjects. These topics should find broad acceptance and scholarly involvement on the part of many readers interested in such generally important areas of acid mine drainage using coal fly ash. Coal fly ash beneficiation has come to assume a very important part of many, if not most, research areas in materials science and numerous other areas of current, scientific research and development in recent times. Therefore, it is imperative for all those involved in these and other research and development areas to keep themselves up-to-date and abreast in the very latest developments in these and other areas of coal fly ash beneficiation.

Segun A. Akinyemi
Department of Geology and Applied Geophysics
Faculty of Science
Ekiti State University
Ado Ekiti, Nigeria

Mugera W. Gitari
Department of Ecology and Resource Management
School of Environmental Science
University of Venda
Thohoyandou, South Africa

Chapter 1

Introductory Chapter: Coal Fly Ash and Its Application for Remediation of Acid Mine Drainage

Mugera Wilson Gitari and Segun Ajayi Akinyemi

Additional information is available at the end of the chapter

http://dx.doi.org/10.5772/intechopen.70711

1. Introduction

The coal mining and other metal extraction industry contributes significantly to the country's economy through export markets of the processed minerals in addition to creation of thousands of jobs in the mining industry. For example, in South Africa, 98% of the country's electricity supply is from coal combustion. Coal mining and coal combustion for electricity generation leaves in its wake various types of waste streams such as acid leachate generating coal discards, acid generating mine tailings, and acid mine drainage (AMD) in mine voids and huge volumes of coal fly ash (CFA) in power utilities. Environmental regulations require these waste streams to be managed and remediated to reduce any negative impact on the environment. Of most concern and increasing interest is the prevention and control of acid mine drainage (AMD) and its treatment. Acid mine drainage has the potential to impact negatively on aquatic ecosystems and groundwater availability in particular. Acid mine drainage (AMD) is generally regarded as the principal environmental problem caused by the mining of the sulfide ore deposits. In coal combustion in power utilities, huge volumes of coal fly ash (CFA) are produced annually and this is normally stockpiled on land as ash heaps that require management due to possible negative environmental impacts such as generation of leachates with high concentration of toxic metal and non-metal species which could affect the quality of surface and groundwater resources in addition to the generation of particulate pollution.

2. Acid mine drainage generation and negative environmental impacts

Acid mine drainage is generated when sulfide minerals in the mined rock or overburden are exposed to oxygen and water during the mining process. These sulfide minerals include but not limited to pyrite (FeS_2), its dimorph marcasite, and pyrrhotite ($Fe_{1-x}S$). These sulfide minerals undergo bacterially catalyzed oxidation reactions which generated acidity and increased Fe, sulfate, and other toxic metal species concentration in recipient water bodies such as groundwater in mine voids [1] (Eq. (1))

$$FeS_2(s) + 3.75\,O_2 + 3.5\,H_2O \rightarrow Fe\,(OH)_3(s) + 2\,SO_4^{2-} + 4\,H^+ \tag{1}$$

AMD is extremely acidic (as low as pH 2.0) and enriched with iron, manganese, aluminum, sulfate, and metal species such as lead, mercury, cadmium, and zinc depending on the geology of the mined rock [2]. During active mining operations, the groundwater is normally pumped out to maintain water table below mining levels but once mining stops pumping stops and the mine voids fill with acid mine drainage and eventually decant to the surface. This mine water requires management, which involves storage and treatment to the specific country's disposal guidelines before disposal to surface water bodies.

Influx of acid mine drainage (AMD) into streams can degrade both aquatic habitat and water quality, often producing an environment devoid of most aquatic life. The extent of impact on aquatic ecosystems depends on a variety of factors which includes frequency of influx, volume, and chemistry of the AMD, and the buffering capacity of the receiving aquatic ecosystem [3]. Drainage from underground mines, surface mines, and refuse piles is the oldest and most chronic industrial pollution associated with coal mining. Various impacts of AMD include quality impacts on groundwater, corrosion of water supply infrastructure, and other manmade structures such as pipes, bridges, dams, and pumps. Acid mine drainage can also be toxic to vegetation; however, toxicity depends on discharged volume, pH, total acidity, and concentration of dissolved chemical species. pH is the most critical with respect to aquatic life. Smothering of stream beds from precipitated metal compounds is a common phenomenon in AMD-impacted streams [4, 5]. Acid mine drainage can also cause reduction in diversity and total number of micro-invertebrates and changes in community structure.

3. Coal fly ash generation and its physicochemical properties

Coal fly ash (CFA) is a by-product obtained during the combustion of coal in coal-burning power generation plants. As demand for cheaper electricity increases, huge volumes of coal fly ash will be generated that will require disposal and management. The worldwide annual production of CFA stands at 780 Mt [6] while 415 Mt or 53% is beneficially utilized, but utilization varies across countries. Japan has the highest utilization rate at 96.4%, Europe 90.9%, China 67%, and Middle East and Africa 10.6% [6]. The rest of the CFA is stockpiled on land or slurried to ash dams [7, 8]. South African Bureau of Standards (SABS) [9] defines fly ash as the powdery residue obtained by separation of the solids from the flue gases of furnaces fired with pulverized coal. Coal fly ash consists of many small (0.01–100 μm diameter) glass-like particles of a generally spherical character. The fineness of fly ash particles depends largely on the combustion temperature, the grinding size of introduced coal, and whether the resultant particle is spherical or irregular. The physicochemical and mineralogical properties of CFA depend on the composition of the parent coal, the conditions during coal combustion (temperature, air/fuel ratio, coal pulverization size, and rate of combustion), the efficiency of emission control devices, the storage and handling of the by-products, and the climate [10, 11].

Coal fly ash tends to accumulate toxic chemical species at high temperatures involved during its generation [10, 12] and is considered an environmental hazard in some countries. Anionic species (Cl^-, SO_4^{2-}), oxy-anionic species (Se, As, Mo, B, and Cr), and cationic species (Al, Fe, Na, K, Ca, Sr, Ba, Zn, Cu, Cd, and Mg) are leached from the ash heaps by the wastewater derived from the ash slurry or by subsequent infiltration by rain upon disposal [8, 10, 13]. This may be of environmental concern due to possible contamination of surface and groundwater in disposal sites and limits beneficial application of CFA. The pH of a fly ash suspension, for example, in water can vary depending on the S content of the parent coal [14]. Fly ashes derived from anthracite coals are generally high in S and produce acidic fly ashes while fly ashes derived from lignite coals are low in S but high in Ca and produce alkaline ashes [15, 16]. South African coal is sub-bituminous and generates fly ash that is characterized by low Fe content. The aqueous extracts of this high Ca coal fly ash are strongly alkaline (pH 12–12.5) due to the free lime content [2, 17]. Mineralogical analysis indicates coal fly ash to be mainly aluminosilicate which forms the basis of its utilization in the synthesis of geopolymers [18], zeolitic adsorbents for water treatment [19]. X-ray diffraction also indicates that CFA has free lime content ranging from 1 to 40% depending on the coal source [6]. This free lime content is the basis of fly ash utilization in acid mine drainage treatment [2] and remediation of acidic mine spoils and control of acid mine drainage generation in sulfidic mine tailings [11].

4. Acid mine drainage treatment using coal fly ash

Acid mine drainage is characterized by high acidity (pH 2–4) and often contains high concentrations of chemical species such as Fe, Mn, and Al and anionic species such as SO_4 in addition to elements like Zn, Co, Pb, Cr, and Cu, in trace concentrations which necessitate these waters to be treated before release. Management of mine water pollution demands a range of active and passive remediation engineering technologies to minimize its impact on ground and surface waters which can incur significant expense [20]. Consequently, mining companies are in constant search for innovative and economically viable treatment technologies. Traditionally, the remediation of AMD has been carried out through a range of active and passive technologies. Active treatment technologies involve the use of alkaline reagents such as limestone and lime [21, 22], and magnesite [23]. Some of the limitations of limestone treatment processes include armoring of limestone particles by amorphous ferric hydroxide precipitates which reduces the efficiency of treatment process and attains a maximum pH of 7 which leaves species such as Mg in solution [24]. A limitation of lime is its high cost [24]. Passive treatment systems involve a combination of alkaline reagents and utilization of natural biochemical processes in artificially constructed wetlands, ponds, and alkaline-generating drains [25–27]. Innovative acid mine drainage treatment using unconventional alkaline agents such as slag has been evaluated [28, 29]. The steel slags on contact with the AMD increased the pH to circumneutral levels and reduced the levels of most major inorganic contaminants such as

Al, Fe, Ti, Ni, Be, and Cu. However, the slag was observed to leach chemical species in the reaction mixture leading to an increased concentration of Ba, V, Mn, Cr, As, Ag, and Se in the leachates. This would be a major drawback of employing the steel slags in AMD treatment due to the secondary contamination of the product water.

Several research studies have reported on the application of coal fly ash for the amendment of acidic coal mine spoils [30] and of acidic soils [31]. These applications were motivated by the alkaline nature of the coal fly ash. Other researchers have reported on the ability of the coal fly ash to remove metal species from aqueous solutions [32–34].

Due to the pressure on the mining companies to reduce the cost of acid mine drainage treatment, they are constantly looking for cheaper treatment agents for AMD. On the other hand, most coal combustion power utilities are constantly looking for large volumes of beneficial coal fly ash generated by their coal combustion process. In most instances, the coal mines generating AMD are located close to the power utilities producing coal fly ash. This has motivated researchers to look at the possibility of utilization of coal fly ash as liming agent for AMD. Coal fly ash contains free lime content that can be used in the neutralization of AMD leading to an increase of pH and precipitation of the metal ion contaminants as insoluble hydroxides, oxyhydroxides, or oxyhydroxysulfates [17, 35].

Coal fly ash has also been beneficiated into zeolitic materials that are used as adsorbents for metal species in AMD effluents. These zeolites possess adsorptive properties and provide a combination of ion exchange and molecular sieve properties [36, 37]. Results of the treatment of AMD with CFA zeolitic product indicated an increased pH and a decreased metal concentration. These materials have advantage over traditional liming materials such as lime or limestone since they are cheaply produced and contribute to sustainable management of coal fly ash.

Although coal fly ash will continue to attract researchers working on cheaper options for the treatment of AMD in both active and passive systems, it is important to note some of the shortcomings and strengths of its application for AMD effluent treatment. Several authors [2, 17, 35] observe that coal fly ash can effectively increase the pH of AMD and decrease its metal species concentration leading to much cleaner product water. The process was observed to be effective for the treatment of acidic AMD [2]. However, the application of CFA for AMD treatment leads to the release of Na, Cl, K, Mg, B, Sr, Ba, and Mo leading to an increased salinity of product water. The chemistry of the coal fly ash and AMD being treated will be a significant determinant factor on the success of the treatment process in addition to the contact time employed. The authors further observed that the treatment process will be combination specific, meaning that different CFA:AMD combination ratios will give product water with varying quality.

5. Conclusions

Coal fly ash can be used effectively to treat AMD. However, the process has its strengths and weaknesses. The treatment technology will be dependent on the chemical properties of the coal fly ash and the AMD effluents being treatment. This means that the treatment process has to be optimized for each coal fly ash/AMD combination for effective results. The product water in this treatment process will require secondary treatment such as reverse osmosis to remove the increased salinity of the product water. Most countries have environmental legislation that classify coal fly ash as a hazardous material, hence reducing its development and utilization as a beneficial material. However, there is still a lot to be learned in terms of the application of coal fly ash and its products for remediation of acid mine waters.

Author details

Mugera Wilson Gitari[1]* and Segun Ajayi Akinyemi[2]

*Address all correspondence to: mugera.gitari@univen.ac.za

1 Environmental Remediation and Water Chemistry Group, School of Environmental Sciences, University of Venda, Thohoyandou, South Africa

2 Department of Geology, Faculty of Science, Ekiti State University, Ado Ekiti, Nigeria

References

[1] Nordstrom SK. The effect of sulfate on aluminum concentrations in natural waters: Some stability relations in the system Al_2O_3-SO_3-H_2O at 298°K. Geochimica et Cosmochimica Acta. 1982;**46**:681-692. DOI: https://doi.org/10.1016/0016-7037(82)90168-5

[2] Gitari MW, Petrik LF, Etchebers O, Key DL, Iwuoha E, Okujeni C. Treatment of acid mine drainage with fly ash: Removal of major contaminants and trace elements. Journal of Environmental Science Health-Part A. 2006;**A41**:1729-1747. DOI: 10.1080/10934520600754425

[3] Kimmel WG. The impact of acid mine drainage on the stream ecosystem. In Pennsylvania Coal: Resources, Technology Utilization, (S.K. Majumdar and W.W. Miller, eds), The pa. Acad. Sci. Publ. 1983;424-437. ISBN-13: 978-0960667017

[4] Parsons JD. The effects of acid strip-mine effluents on the ecology of a stream. Archiv für Hydrobiologie. 1968;**65**:25-50

[5] Warren CJ, Dudas MJ. Formation of secondary minerals in artificially weathered fly ash. Journal of Environmental Quality. 1985;**14**:405-410. DOI: 10.2134/jeq1985.00472425001400030019x

[6] Craig H, Hans-Joachim F, Anne W. Coal combustion products: A global perspective. In: World of Coal Ash (WOCA) Conference; April 22-25, Lexington, KY, USA 2013. Available from: http:/www.flyash.info/

[7] Abbott DE, Essington ME, Mullen ND, Ammons JT. Fly ash and lime-stabilized biosolid mixtures in mine spoil reclamation: Simulated weathering. Journal of Environmental Quality. 2001;**30**:608-616. DOI: 200110.2134/jeq2001.302608x

[8] Mattigod SV, Rai D, Eary LE, Ainsworth CC. Geochemical factors controlling the mobilization of inorganic constituents from fossil fuel residues: Review of the major elements. Journal of Environmental Quality. 1990;**19**:188-201. DOI: 10.2134/jeq1990.00472425001900020004x

[9] South African National Standards (SANS 50413-1/EN 450-1). Fly Ash for Concrete—Part 1: Definition, Specifications and Conformity Criteria. 2015. ISBN 978-0-626-32353-0

[10] Eary LE, Dhanpat R, Mattigod SV, Ainsworth CC. Geochemical factors controlling the mobilization of inorganic constituents from fossil fuel combustion residues: II. Review of the minor elements. Journal of Environmental Quality. 1990;**19**:202-214. DOI: 10.2134/jeq1990.00472425001900020005x

[11] Xenidis A, Evangelia M, Ioannis P. Potential use of lignite fly ash for the control of acid generation from sulphidic wastes. Waste Management. 2002;**22**:631-641. DOI: http://dx.doi.org/10.1016/S0956-053X(01)00053-8

[12] Spears DA, Lee S. Geochemistry of Leachates from Coal Ash. Geological Society, London Special Publications. 2004;**236**:619-639. January 1, 2004. DOI: https://doi.org/10.1144/GSL.SP.2004.236.01.35

[13] Plank CO, Martens DC. Boron availability as influenced by application of fly ash to soil. Soil Science Society of America, Proceedings. 1974;**38**:974-977. DOI: 10.2136/sssaj1974.03615995003800060038x

[14] Barry SR, Daniels WL, Jackson ML. Evaluation of leachate quality from co-disposed coal fly ash and coal refuse. Journal of Environmental Quality. 1997;**26**:1417-1424. DOI: 10.2134/jeq1997.00472425002600050031x

[15] Furr AK, Parkinson TF, Hinrichs RA, Van Campen DR, Bache CA, Gutenmann WH, St. John RE Jr, Pakkala IS, Lisk DJ. National survey of elements and radioactivity in fly ashes: Absorption of elements by cabbage grown in fly ash soil mixtures. Environmental Science & Technology. 1977;**11**:1194-1201 ISSN: 0013-936X

[16] Page AL, Elseewi AA, Straughan I. Physical and chemical properties of fly ash from coal-fired power plants with reference to environmental impacts. Residue Reviews. 1979;**71**:83-120 978-1-4612-6185-8_2

[17] Gitari WM, Petrik LF, Etchebers O, Key DL, Iwuoha E, Okujeni C. Passive neutralisation of acid mine drainage by fly ash and its derivatives: A column leaching study. Fuel. 2008;**87**:1637-1650. DOI: https://doi.org/10.1016/j.fuel.2007.08.025

[18] Nyale SM, Babajide OO, Birch GD, Böke N, Petrik LF. Synthesis and characterization of coal fly ash-based foamed geopolymer. Procedia Environmental Sciences. 2013;**18**:722-730. DOI: https://doi.org/10.1016/j.proenv.2013.04.098

[19] Musyoka NM, Petrik LF, Gitari WM, Balfour G, Hums E. Optimization of hydrothermal synthesis of pure phase zeolite Na-P1 from South African coal fly ashes. Journal of Environmental Science and Health, Part A. 2012;**47**:337-350. DOI: 10.1080/10934529.2012.645779

[20] Younger PL, Banwart SA, Hedin RS. Mine water: Hydrology, pollution, remediation. In: Chapter Two: Mine Water Chemistry. Dordrecht: Kluwer Academic Publishers; 2002. p. 65-126 978-94-010-0610-1

[21] Maree JP, du Plessis P. Neutralisation of acidic effluents with calcium carbonate. Water Science and Technology. 1994;**29**:285-296. DOI: http://wst.iwaponline.com/content/29/9/285

[22] Geldenhuys AJ, Maree JP, De Beer M, Hlabela P. An integrated limestone/lime process for partial sulphate removal. South African Institute of Mining and Metallurgy (SAIMM). 2001;**103**:345-371 ISSN 2411-9717

[23] Masindi V, Gitari WM, Tutu H, De Beer M. Passive remediation of acid mine drainage using cryptocrystalline magnesite: A batch experimental and geochemical modelling approach. Water SA. 2015;**41**:677-682. DOI: http://dx.doi.org/10.4314/wsa.v41i5.10

[24] Maree JP, Van Tonder GJ, Millard P. Underground Neutralization of Mine Water with Limestone. Pretoria, South Africa: Water Research Commission, Report No. 609/1/96; 1996. ISBN: 1-86845-241-7

[25] Cravotta III AC, Trahan MK. Limestone drains to increase pH and remove dissolved metals from acidic mine drainage. Applied Geochemistry. 1999;**14**:581-606. DOI: http://dx.doi.org/10.1016/S0883-2927(98)00066-3

[26] Hedin RS, Watzlaf GR. The effects of anoxic limestone drains on mine water chemistry. In: Proceedings of the International Land Reclamation and Mine Drainage Conference; Pittsburgh. United States of America; 1994. pp. 185-194. DOI: 10.21000/JASMR94010185

[27] Ziemkiewicz PF, Skousen JG, Brant DL, Sterner PL, Lovett RJ. Acid mine drainage treatment with armoured limestone in open limestone channels. Journal of Environmental Quality. 1997;**26**:1017-1024. DOI: 10.2134/jeq1997.00472425002600040013x

[28] Simmons J, Ziemkiewicz P, Courtney Black D. Use of steel slag leach pads for the treatment of Acid Mine Drainage. Mine Water and the Environment. 2002;**21**:91-99. DOI: https://doi.org/10.1007/s102300200024

[29] Zvimba JN, Siyakatshana N, Mathye M. Passive neutralization of acid mine drainage using basic oxygen furnace slag as neutralization material: Experimental and modeling. Water Science Technology. 2017;**75**:5-6. DOI: 10.2166/wst.2016.579

[30] Bhumbla DK, Keefer RF, Singh RN. Selenium uptake by alfalfa and wheat grown on a mine spoil reclaimed with fly ash. In: Proceedings of Mine Drainage and Surface Mine Reclamation Conference, Washington DC, USA, 1998. pp. 15-21. DOI: 10.21000/JASMR88020015

[31] Phung HT, Lund LJ, Page AL, Bradford GR. Trace elements in fly ash and their release in water and treated soils. Journal of Environmental Quality. 1979;**8**:171-175. DOI: 10.2134/jeq1979.00472425000800020007x

[32] Erol M, Kucukbayrak S, Ersoy-Mericboyu A, Ulubas T. Removal of Cu^{2+} and Pb^{2+} in aqueous solutions by fly ash. Energy Conservation Management. 2005;**46**:1319-1331. DOI: http://dx.doi.org/10.1016/j.enconman.2004.06.033

[33] Panday KK, Prasad G, Singh VN. Copper (ii) removal from aqueous solutions by fly ash. Water Research. 1985;**19**:869-873. DOI: https://doi.org/10.1016/0043-1354(85)90145-9

[34] Doye I, Duchesne J. Neutralization of acid mine drainage with alkaline industrial residues: Laboratory investigation using batch-leaching tests. Applied Geochemistry. 2003;**18**:1197-1213. DOI: http://dx.doi.org/10.1016/S0883-2927(02)00246-9

[35] Madzivire G, Gitari WM, Kumar Vadapalli VR, Ojumu TV, Petrik LF. Fate of sulphate removed during the treatment of circumneutral mine water and acid mine drainage with coal fly ash: Modelling and experimental approach. Minerals Engineering. 2011;**24**:1467-1477. DOI: https://doi.org/10.1016/j.mineng.2011.07.009

[36] Moreno N, Querol X, Ayora C. Utilization of zeolites synthesized from coal fly ash for the purification of acid mine waters. Environmental Science and Technology. 2001;**35**:3526-3534. DOI: 10.1021/es0002924

[37] Fungaro DA, Izidiro JDC. Remediation of acid mine drainage using zeolites synthesized from coal flyash. Química Nova. 2006;**29**:735-740. DOI: http://dx.doi.org/10.1590/S0100-40422006000400019

Chapter 2

Metal Adsorption by Coal Fly Ash: The Role of Nano-sized Materials

Anita Etale, Nikita T. Tavengwa and
Vusumzi E. Pakade

Additional information is available at the end of the chapter

http://dx.doi.org/10.5772/intechopen.69426

Abstract

Nano-sized particles (1–100 nm) comprise a considerable fraction of coal fly ash (CFA). They are unique due to their large surface area and higher reactivity compared to larger CFA particles. As they are formed by nucleation of volatilised elements or through chemical reactions, nano-scale CFA particles have been shown to take up greater quantities of elemental ions and bind them more strongly than larger particles, diminishing the fraction of desorbed ions. Despite this and the large volume of literature on acid mine drainage (AMD) treatment using CFA, little is described about the specific role of nanoparticles in this process. This chapter therefore sets out to highlight this, beginning by delineating nanoparticle characteristics that make them good adsorbents followed by details of their formation and action in metal adsorption.

Keywords: nanoparticles, acid mine drainage, remediation, mining, water pollution

1. Introduction

Coal accounts for up to 29% of energy production worldwide, generating millions of tons of incombustible residue known as coal fly ash (CFA) in the process [1]. The colour of CFA varies from tan to grey to black depending on the amount of unburnt carbon. It consists of fine spherical particles that may be either hollow or solid and have surface areas as high as 1000 $m^2\ kg^{-1}$ [2].

The chemical properties of CFA are primarily influenced by the type of coal burnt, that is, bituminous, sub-bituminous, anthracite or lignite. Anthracite has the highest carbon content, hence also has the highest calorific value of all four types. It however, costs two to three times more than regular coal and is therefore not typically used in commercial boilers. Bituminous coal contains mostly silica, alumina, iron oxide and calcium while lignite and sub-bituminous coal contain less silica and iron oxides but higher calcium and magnesium oxide concentrations (**Table 1**). CFA is also designated based on the concentrations of these four major minerals. Ash with more than 70 wt% silica, alumina and iron oxides and low in lime (1–12%) is designated class F while ash with 50–70 wt% silica, alumina and iron oxides and high lime content (30–40%) is class C [2]. The former is usually produced from combustion of bituminous and anthracite coal, and the latter from lignite and sub-bituminous coals.

Although the greater share of CFA generated after coal combustion is deposited in waste heaps, it is also widely used in road construction, as a raw material in cement manufacture, for the sequestration of toxic gases including SO_2, NO_2 and Hg and, more recently, as a source of catalysts and metals [3, 4].

The formation of mine drainage laden with toxic concentrations of metals and other ions is a common legacy of base metal and coal mining [5, 6]. Acid mine drainage (AMD) is formed via a cascade of reactions when ferrous sulphide phases, for example, pyrite are exposed to oxygenated water, leading to the formation of ferric ion. Ferric ion then subsequently acts as an additional oxidising agent for pyrite, making the reaction self-perpetuating as long as the pH remains below 3.5 although other factors, including temperature and microorganisms, also play an important role [7]. AMD poses significant environmental and ecological challenges to receiving water bodies as well as to human populations that depend on these water resources for consumption [8]. There have therefore been numerous attempts at investigating the use of various materials including CFA which is otherwise a waste product, for the treatment of AMD. Two main characteristics stand out in favour of CFA: (i) its alkaline nature that counters AMD acidity and (ii) its capacity for adsorption/co-precipitation of elemental ions [9, 10]. The synthesis of other adsorbents from CFA, for example, zeolites [11–16], modified mesoporous silica [17] and a number of other composite materials [18–20] have also been widely studied.

Metal removal by CFA and its derivatives varied, following different mechanisms including ion exchange, precipitation as well as chemical and physical adsorption as the following

Components	Bituminous (%)	Sub-bituminous (%)	Lignite (%)
Silica	20–60	40–60	15–45
Alumina	5–35	20–30	10–25
Ferric oxide	10–40	4–10	4–15
Calcium oxide	1–12	5–30	15–40
Magnesium oxide	0–5	1–6	3–10
Sulphite	0–4	0–2	0–10
Sodium oxide	0–4	0–2	0–6
Potassium oxide	0–3	0–4	0–4

Table 1. Chemical composition of fly ash from the combustion of bituminous, sub-bituminous and lignite coal [2].

selected examples illustrate. Belviso et al. [12] investigated the removal of Mn^{2+} by raw CFA from Brindisi, Italy, and a zeolite synthesised from the same fly ash. They found that adsorption was rapid, attaining equilibrium in 5 minutes and that samples filtered with untreated CFA which contained higher Ca^{2+} and Mg^{2+} concentrations, for example, up to 340 mg L^{-1} versus 22 mg L^{-1} of Ca in CFA and zeolite-treated samples, respectively. In addition, although CFA resulted in slightly greater removal in both batch and column tests, >97% of Mn^{2+} was removed from solution by the end of exposure times. With pH_{PZC} of 12.2 and 12, adsorption of Mn^{2+} to either adsorbents was unlikely as all were positively charged. Since Mn^{2+} precipitates above pH 9 as $Mn(OH)_2$, the high solution pH induced by adsorbents likely resulted in its precipitation. These high efficiencies were therefore the result of Mn^{2+} precipitation induced by the high pH of the adsorbents. Indeed, as noted by Cho et al. [9], fly ash is strongly alkaline with pH ranging from 10 to 13 when added to water. In contrast to the preceding study, these authors found that their CFA, sourced from Boryug, Korea, had a pH_{PZC} of 2.8. It was therefore primed for adsorption of positively charged metal ions over a wide pH range and indeed, removed >95% of Zn^{2+}, Pb^{2+}, Cd^{3+} and Cu^{2+} by pH 8. Further, although removal of Zn was reduced in the presence of Cu^{2+} and Pb^{2+}, that of Cd^{3+} was improved in the presence of Pb^{2+}, and removal of Pb^{2+} itself was unaffected by any of the other metals. In general, adsorption by this CFA was in the order of $Pb^{2+} > Cu^{2+} > Cd^{3+} > Zn^{2+}$. Adsorption equilibrium was attained in 2 hours for Zn and in 3 hours for the rest of the metals, and isotherm data for all four were well fitted by the Freundlich isotherm.

Spherical porous pellets made from Class C CFA from power plant in Greece were used by Papandreou et al. [21] and applied in the removal of Pb^{2+}, Zn^{2+} and Cr^{3+}. The pellets had a Brunauer-Emmett-Teller surface area of 23.41 m^2 g^{-1} and approximately 52% of pore volumes occupied by 3.5–4.5 nm pores. Adsorption was investigated at pH 1–10 and found to be optimal at pH 7, for Pb^{2+} and Cr^{2+} and pH 8 for Zn^{2+}. Electrostatic attraction between the ions and the charged adsorbent surface was determined to be the adsorption mechanism. Adsorption isotherms were best fitted by the Langmuir isotherm and maximum adsorption capacities of the pellets calculated to be 0.22, 0.27and 0.44 mmol g^{-1} for Pb^{2+}, Zn^{2+} and Cr^{3+}, respectively. Zinc adsorption was strongly inhibited in the presence of Pb, a finding similar to that of Cho et al. [9] above, and which implies that these ions sorb to similar sites on both types of CFA. In contrast, Cr^{3+} did not inhibit sorption of either Pb^{2+} or Zn^{2+}.

The relationship between adsorption and temperature is often used as an indicator of whether the process involves chemical or physical bond formation. An inverse relationship, that is, where adsorption decreases with increasing temperature denotes physisorption while a direct relationship, that is, adsorption increasing with increasing temperature denotes chemisorption. Adsorption to the pellets in the study by Papandreou above was found to substantially increase with temperature from 25 to 60°C implying that sorption of Pb^{2+}, Zn^{2+} and Cr^{3+} was via chemical bond formation.

The use of CFA for the adsorption of metal ions from acid mine drainage was investigated by Rios et al. [22]. A class F CFA from the Rugeley power station in the West Midlands of England was applied to the strongly acidic (pH < 2) AMD from the Parys mountain in Anglesley, North Wales. Batch experiments were conducted to determine adsorption of Pb, Cu, Zn, Ni, Cr, Fe,

As and NH_4^+ onto the raw CFA, a natural clinker from the Cerrejón coal deposit in Colombia and synthetic zeolites. Using sorbent/AMD ratios of 0.25 g/20 mL and 1 g/20 mL over a period of 24 hours, pH rose only slightly from 1.96 to 2.66 after 1 hour but no breakthrough to alkaline pH was observed using this adsorbent dosage. Adsorption by CFA seemed to partition into two scenarios. In the first, concentrations of the target ion in the final solution were higher than in the original untreated AMD. These ions included NH_4^+, Cr^{3+}, Cu^{2+} and Zn^{2+}. In the second group, CFA adsorbed more ions than it released into solution and final concentrations were lower; As, Zn, Fe and Pb belonged to this group. This study highlights the challenge of using raw CFA in AMD treatment, that is, that although some elemental ions may be attenuated, the concentrations of others may increase.

Column studies provide the opportunity to assess the long-term applicability of CFA treatment of AMD. Pérez-López et al. [23] used column experiments to determine the efficacy of fly ash in attenuating AMD formation from pyritic sludge. They found that while columns filled with pyrite-rich sludge leached drainage that was acidic (pH ~ 2) and rich in sulphate, iron and other metals, those containing sludge and fly ash produced alkaline leachate low in sulphate and lacking iron or other metals. The alkaline fly ash neutralised AMD acidity, thus decreasing metal solubility and increasing their precipitation within the column. In addition, these precipitates formed coatings on pyrite grains, preventing their exposure to oxidising agents thus retarding AMD generation in a process known as microencapsulation [24].

Gitari et al. [10] used column studies to investigate the suitability of CFA and its derivatives as back-fill materials. They therefore designed experiments to measure the AMD treatment efficacy of (i) fresh CFA, (ii) solid residues prepared from exposing CFA to AMD, (iii) blends of the solid residues with varying amounts of fresh CFA (5, 25 and 40%) and (iv) a blend of solid residues with ordinary portland cement (OPC). They found that dissolution of CaO in fly ash and the solid residues, and hydration reactions in OPC were responsible for the initial high pH of the leachates in the different columns. This alkalinity subsequently controlled metal leaching through the formation of aluminium and iron precipitates and the co-precipitation of other elements with these phases. However, the leached fractions of most ions increased with increasing CFA concentrations in the blends, for example, Cr(III) concentrations went from 4.97 to 10.27 $\mu mol\ L^{-1}$ and 11.19 $\mu mol\ L^{-1}$ as CFA concentrations in solid residue blends increased from 5 to 25% and then to 40%. Sulphates were removed as gypsum ($CaSO_4$) and as such, their attenuation was highly dependent on Ca^{2+} from dissolution of CaO in CFA and OPC. Attenuation of Fe and Al, on the other hand, was pH dependent since the formation of hydroxide precipitates was greater at neutral and basic pH. Subsequent modelling studies using PHREEQ software showed that Fe concentrations were controlled by goethite and amorphous $Fe(OH)_3$ phases while those of Al were controlled by boehmite, basaluminite and gibbsite [25].

In some cases, for example, where AMD passes through dolomitic rocks, the acidity as well as metal and sulphate concentrations may be attenuated, resulting in neutral mine waters. Despite this, sulphate concentrations often remain considerable due to high initial concentrations [26]. Madzivire et al. [27, 28] investigating the use of CFA for sulphate removal found that attenuation was not significant below pH 10. Nevertheless, 60% of sulphate was removed from the solution at pH above 11 with the addition of gypsum and alumina, and sulphate was removed as ettringite ($Ca_6Al_2(SO_4)_3(OH)_{12}{\cdot}26H_2O$). One drawback, however, was that the use of CFA increased the concentrations of B, K, Cr, Mo and Sr in the treated water necessitating the inclusion of an additional treatment step such as adsorption using zeolites. Nevertheless, these studies provide evidence of the applicability of CFA in AMD treatment both in acid neutralisation and metal removal.

Metal removal from AMD proceeds through various mechanisms including precipitation, oxidation, reduction and adsorption. Adsorption efficiency increases with surface area. The large surface areas of CFA splenospheres and cenospheres therefore significantly influence metal adsorption from AMD. Specifically, since surface areas increase with decreasing particle size, smaller particles, particularly nano-sized ones (1–100 nm), likely play a very important role in adsorption of element ions. In fact, CFA has been shown to contain significant quantities of nanoparticles [29, 30] which are enriched in metal ions [31–36]. Despite this, there is a dearth of information regarding the role of CFA nanoparticles during AMD treatment.

This chapter therefore highlights the key role of nanoparticles in remediation effects of CFA. Using Fe oxy(hydroxides), Ti dioxides and carbon nanotubes phases known to be abundant in CFA [29, 37–39], we illustrate the likely effect of nanoparticle-based processes on observed metal removal efficiencies of CFA.

2. Nanomaterials in CFA

2.1. Properties of nanomaterials

Consider a crystal of iron or titanium oxide. In the bulk of the crystal, Fe or Ti ions are surrounded (coordinated) by six nearest-neighbour oxygen atoms, an arrangement that balances the electric charge of anions and metal atoms. However, when the bulk is divided into smaller particles, new surfaces are created and the spatial configuration of the bulk is lost. This leads to an increase in surface energies which the newly exposed atoms on the surface attempt to dissipate by re-organising, for example, 'dangling' oxygen atoms may form bridged pairs. Such surface reconstructions result in particles with considerably different properties from those of the bulk material [40, 41].

In fact, many of the unique characteristics of nanomaterials (NMs) are the result of size-dependent surface properties. There is evidence that properties, including crystalline habit, magnetism and colour, change with particle size [42]. In addition, the proportion of atoms located at or near particle surfaces increases as particle size decreases. The extra surface charge increases particle reactivity, hence smaller particles are more reactive than bulk counterparts [43, 44]. Brown et al. [45] also showed that the larger number of edge and corner binding sites as well as the wide variety of defects on the surfaces of NMs (**Figure 1**) is responsible for increased reactivity at the nano-scale. Adsorption of ions from surrounding media may be a means to reduce surface energy and regain particle stability. Thus, NMs are excellent adsorbents due to their large surface areas and abundant, highly reactive surface sorption sites.

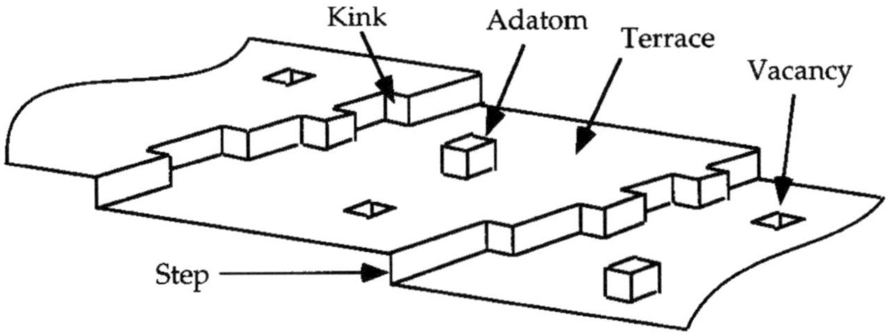

Figure 1. A model of possible defects on single-crystal surface [45].

2.2. Evidence of nanomaterials in CFA

Advancements in various microscopic and spectroscopic techniques have facilitated the identification of nano-sized materials in CFA. By combining transmission and scanning electron microscopy, as well as atomic force microscopy and X-ray diffraction (XRD) for instance, Dwivedi et al. [46] found that CFA from Harduaganj in Aligard, India, contained spherical particles with sizes ranging from 11 to 25 nm. In another study, high resolution transmission electron microscopy (HR-TEM) and energy-dispersive X-ray spectroscopy (EDS) revealed the presence of 10–100 nm hematite and yavapaiite ($KFe(SO_4)_2$) nanoparticles in CFA from a power plant in Santa Catarina state in Brazil [34].

Ribeiro et al. [32] coupled EDS with field-scanning electron microscopy (FE-SEM) and HR-TEM to characterise CFA from an anthracite-burning thermal power plant in, Porto, Portugal. Besides the larger cenospheres (~ 100 μm), ultrafine glassy spheres (~ 100 nm) as well as carbonaceous nanospheres (~ 50 nm) and nanotubes (0.1–32 nm in diameter and 8–59 μm in length) were identified in the CFA (**Figure 2**).

Fe was present as maghemite, hematite and magnetite or as sulphates, that is, schwertmannite and jarosite and titanium as anatase and rutile. Quartz nanoparticles were also identified occurring discretely or embedded in amorphous phases, for example, glass and nano-spherules rich in Al, Si, Ti, K, Mg and Fe were frequently observed embedded in carbonaceous matrices. Indeed, these inorganic phases are often found encapsulated by carbonaceous matter although Wilcox et al. [47] showed that this encapsulation may be a factor of the amount of carbon in feed coal. Comparing CFA from pure high sulphur-high Fe bituminous coal and a blend consisting of 70% of this high sulphur-high Fe coal and 30% of a low sulphur-high CaO coal, these authors found that unlike CFA from the pure coal, inorganic particles in CFA from the blend did not contain any carbon on their surface (**Figure 3**). The composition of CFA particles therefore depends, to some extent, on the nature of feed coal burned; with the amount of carbon rather than the rank of coal determining coal deposition on CFA particles.

As a final example, let us consider the ultrafines (particles < 100 nm) in CFA from US coals examined by Chen et al. [48]. These nanoparticles were found to have distinctly different morphologies, compositions and microstructures (**Figure 4**). In addition, CFA from a bituminous

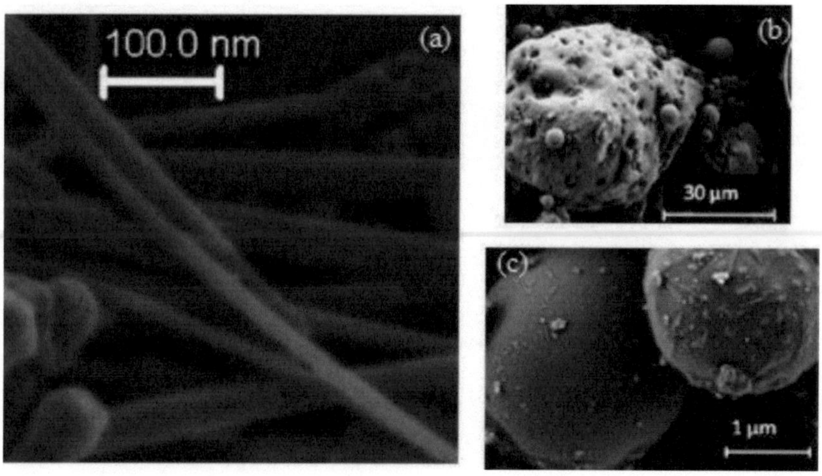

Figure 2. Carbon nanotubes (a) and carbonaceous spheres (b) and (c) in CFA from anthracite coal [32].

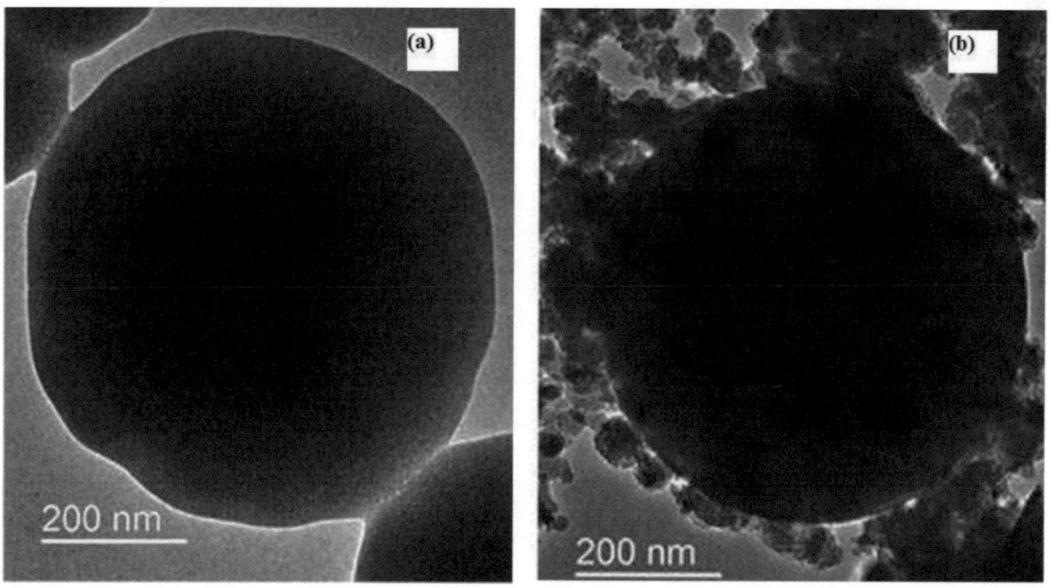

Figure 3. TEM images of (a) particles from the CFA blend. Note the absence of carbon deposits on the surface. (b) CFA particles from combustion of pure high sulphur-high Fe bituminous coal [47].

coal was rich in crystalline phases of Fe, Ti and Al as small as 10 nm while a low-rank CFA was rich in aggregates of alkaline-earth elements in the form of sulphates, silicates and phosphates.

Nano-scale unburned carbon was present as soot aggregates ranging in size from 100 to 1000 although aggregated smaller than 100 nm were also present. Their morphology was similar to that of soot aggregates from diesel engine exhausts. Importantly, these aggregates are made up of 20–50 nm primary particles arranged in an onion-like structure (**Figure 5**). Their formation,

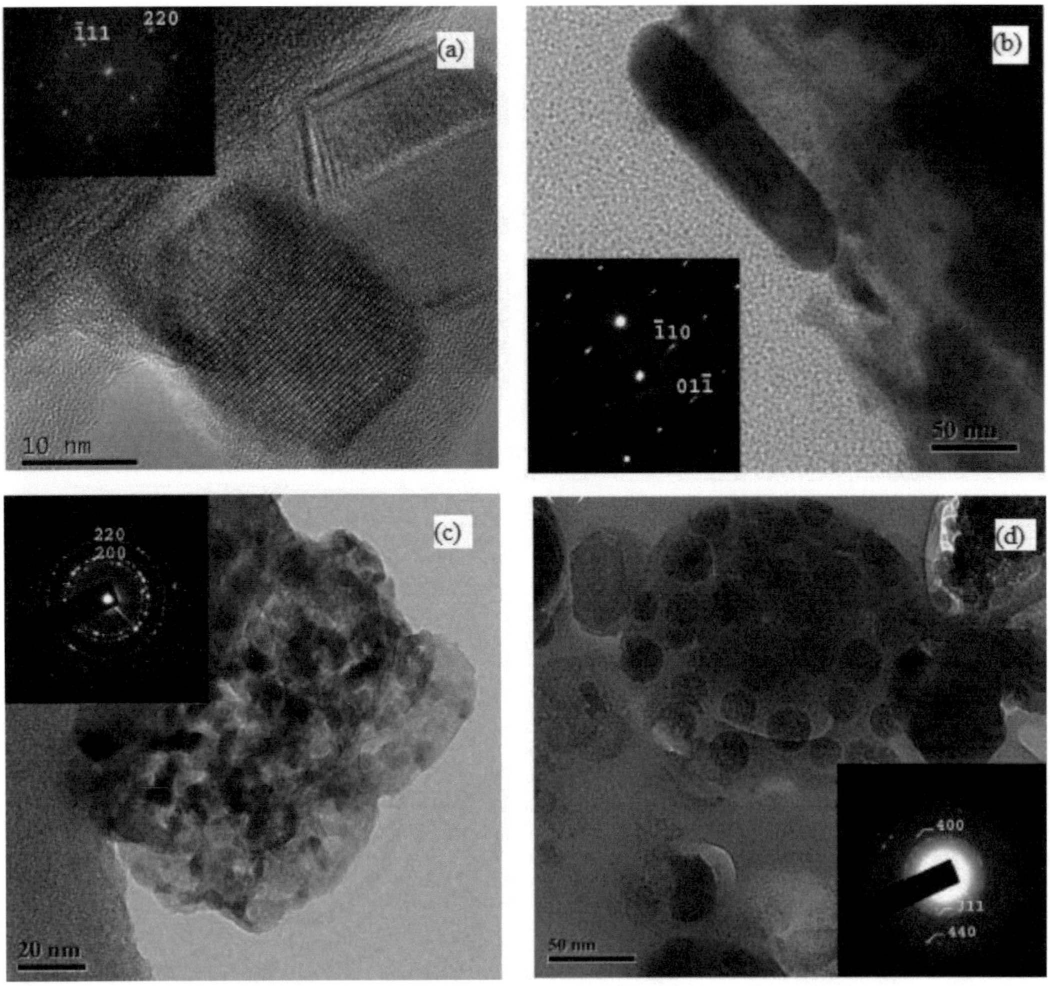

Figure 4. (a) Magnetite nanocrystalline particles, (b) a rod-like rutile particle, (c) a lime particle containing 5–10 nm nanocrystallites and (d) a cluster of $FeAl_2O_4$ particles on an amorphous silica sphere. Inserts in (a) and (b) are fast Fourier transforms (FFT) of the images and those in (c) and (d) are selected area electron diffraction (SAED) patterns [48].

as we will see shortly, likely involves sudden quenching and homogenous condensation of refractory carbon-rich species in the combustion atmosphere.

2.3. Mechanisms of nanomaterial formation in CFA

Having confirmed their presence, let us now briefly consider the mechanisms that lead to the formation of nano-sized materials in CFA. While several mechanisms are responsible for particulate formation in CFA, the formation of nanoparticles is dominated by (i) the release of organically bound metals during coal volatilisation and (ii) vaporisation and subsequent condensation and aggregation of such elements [30, 49]. Nanoparticles in CFA may be categorised as either organic or inorganic and are formed through two separate pathways as illustrated in **Figure 6**.

http://dx.doi.org/10.5772/intechopen.69426

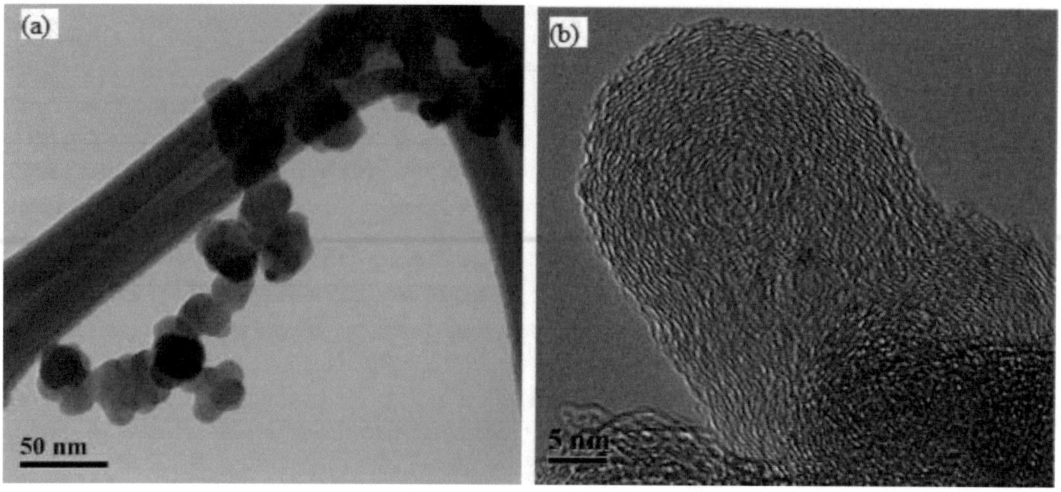

Figure 5. (a) Coal fly ash soot aggregates with chain-like branching structure, (b) concentric stacking of graphitic layers making up soot particles [48].

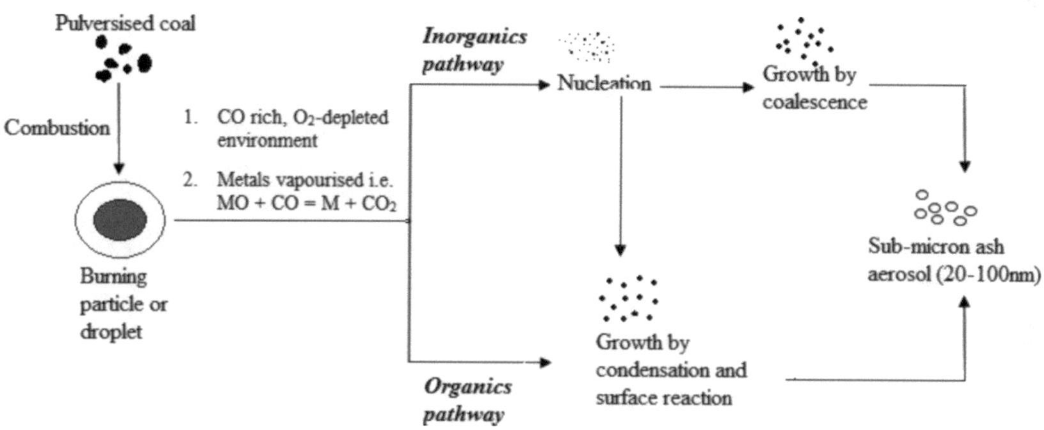

Figure 6. Formation of nanomaterials in coal fly ash.

The formation of inorganic particles occurs through nucleation, also known as homogeneous condensation of flame-volatilised species [50–54], for example, SiO_2, CaO and Fe_2O_3. The main features of this process were outlined by Nelson [51] thus:

(i) The highly reducing conditions existing within coal chars during combustion reduce refractory oxides to more volatile sub-oxides or elements, for example, for Si:

$$SiO_2(s) + CO\,(g) \rightarrow SiO\,(g) + CO_2(g) \tag{1}$$

(ii) As the volatile species move away from the particle to the bulk gas, they are re-oxidised due to the higher O_2 concentrations

$$SiO(g) + O_2(g) \rightarrow SiO_2(g) + O(g) \tag{2}$$

(iii) Provided the vapour pressure of the oxide exceeds the saturated vapour pressure, spontaneous condensation will occur and nuclei will form. Particle growth then proceeds via coalescence. Particle growth may also be the result of heterogeneous condensation, that is, condensation of other species upon these particles. This happens mostly from species that vaporise but whose concentrations are not sufficient for homogenous condensation.

Organic particles, or soot, form under fuel-rich conditions where hydrocarbon fragments have greater chances of colliding and growing instead of being oxidised to CO, CO_2, H_2 and H_2O. Their formation involves chemical reactions that produce clusters of increasing molecular weight to the point where these clusters are visible as particles [30]. The formation of the first aromatic ring, usually benzene, is considered one of the critical steps in soot formation. Benzene rings then go on to form polycyclic aromatic hydrocarbons which increase in molecular weight to soot particles (**Figure 7**).

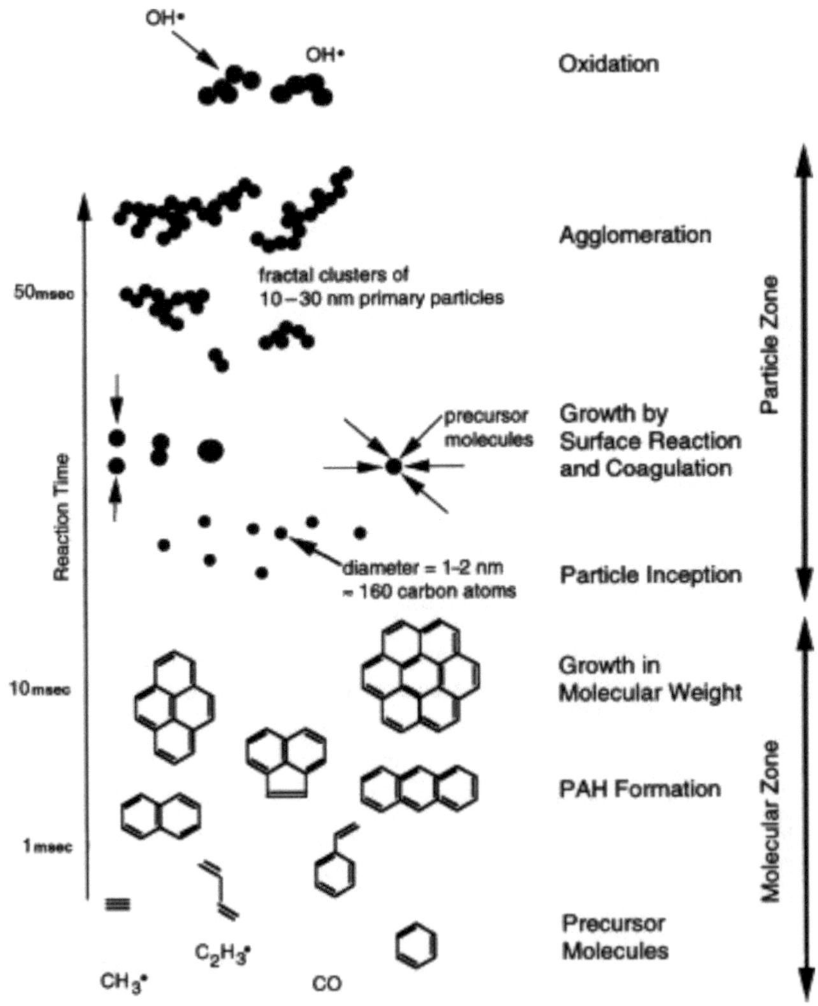

Figure 7. Chemical and physical processes involved in the formation of soot particles [30].

Other factors influencing formation of both organic and inorganic particles include (i) temperature [39, 55], (ii) gas composition [56–59] and coal rank [47, 58, 60]. Particle formation is enhanced at higher temperatures due to increased mineral vaporisation and environments with low oxygen to fuel stoichiometry produce less ultrafines [57]. In addition, Wen et al. [61] recently reported that combustion under N_2 resulted in production of more particulates than combustion in CO_2. The effect of coal rank, on the other hand, is multifarious and can be summarised as follows:

(i) The higher char reactivity of lower rank coals favours the formation of nanoparticles because of the higher combustion temperatures of these chars [62];

(ii) The reducing atmosphere produced by carbon combustion at high temperatures increases ash vaporisation [59];

(iii) Low-rank coals have high concentrations of alkaline-earth metals which supress the vaporisation of silica by reacting with it to form less volatile silicates. This results in nanoparticles from combustion of lignite being mostly MgO and CaO, and SiO_2 nanoparticles being significant only in bituminous coal [59].

3. Metal binding by nanomaterials and larger CFA particles

CFA contains nanomaterials of various parent minerals and elements. These include (i) iron oxides which comprise a large fraction of the inorganic nanoparticles in CFA occurring as pyrite, pyrrhotite, magnetite, hematite, jarosite, goethite and yavapaiite [32, 34, 63–65] and (ii) carbon nanomaterials, for example, single- or multi-walled nanotubes [29, 32, 46, 47].

Both organic and inorganic nanoparticles in fly ash are associated with higher concentrations of metal ions in CFA than larger particles. In fact, it has been known since the work of Davison et al. [52] that element enrichment in CFA is inversely related to particle size. These authors found that concentrations of Pb, Ti, Sb, Cd, Se, As, Ni, Cr and Zn increased with decreasing CFA particle size and that particles with aerodynamic diameters >40 μm, for example, had Pb loadings up to 90 $\mu g\ g^{-1}$ while those with diameters below 5 μm had loadings of up to 980 $\mu g\ g^{-1}$. Positing at first that this observation was due to the fragmentation of pyritic inclusions in the coal, they later inclined to the homogenous and heterogeneous condensation hypotheses explained above. Later, Linak et al. [66] also reported up to 50 times enrichment of S, Cl, Na, K, V and P in ultrafines; confirming that the smaller the particles, the higher the concentrations of metals sorbed onto their surfaces.

Yi et al. [39] studied the relative distributions of trace elements in bottom ash, fly ash and flue gas. They found that except Hg whose concentration was highest in flue gas, all other elements (As, Cd, Cr, Cu, Al, V, Zn, Mn and Fe) were enriched in fly ash relative to bottom ash (**Figure 8**). Similar results were reported for various Indian fly ashes [37, 38, 67]. Hg, although occurring in high concentrations in flue gases, is also bound to a considerable extent to inorganic fractions as we will see later, and to the organic fraction of CFA by carboxylic acid functional groups [29].

For some time, this phenomenon (element enrichment in nanoparticles) was thought of as being purely the result of the greater surface areas available for adsorption in smaller particles

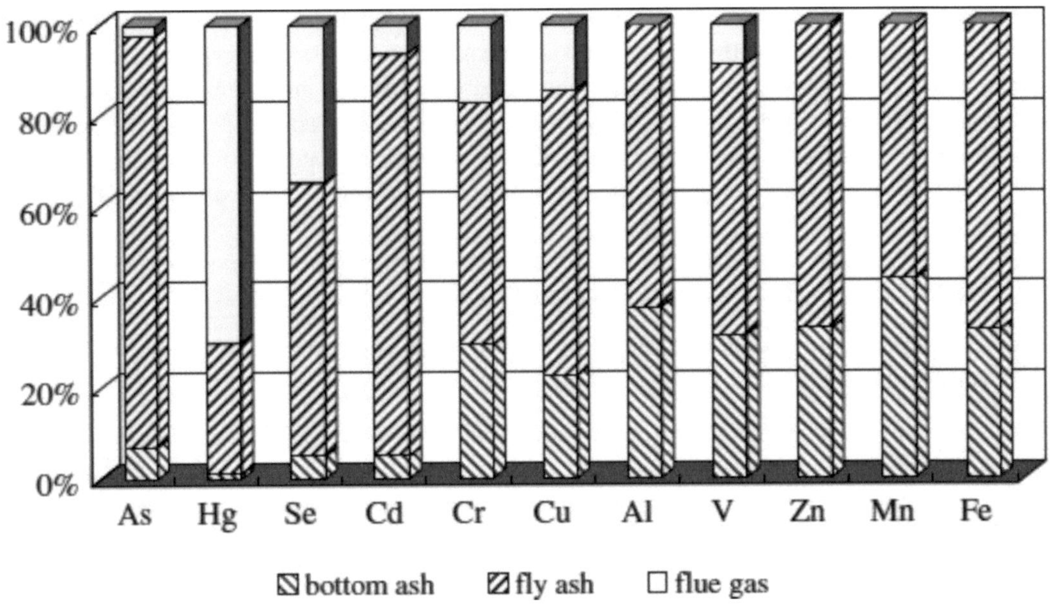

Figure 8. Relative distributions of trace elements in bottom ash, fly ash and flue gas from a coal boiler [39].

[52]. However, later geochemical investigations showed that surface features of these particles played a significant role. Nanoparticle binding of metal ions has been shown to involve faster kinetics [68–70], speciation transformations of the adsorbed species [71] and greater affinity for metal ions [69]. In addition, there is evidence that some of these phenomena, specifically higher reaction rates, are due to surface defects. Junta and Hochella [72] found that in the oxidation of Mn^{2+} to Mn^{3+} at hematite nanoparticle surfaces, steps were the most reactive sites for initiating the adsorption-oxidation reaction. Later, Madden and Hochella [68] found that this oxidation process was up to 1.5 orders of magnitude faster on 7.3 nm particles than on 37 nm particles largely due to changes in the electronic structure of hematite. As its particles get smaller, the Lewis basicity of the surface oxygen increased, decreasing the redox potential of adsorbed Mn^{2+} and making the oxidation reaction more favourable.

The nature of binding sites at the nano-scale has also been invoked to explain changes in metal adsorption. Greater binding of Cu^{2+} ions to 7 nm hematite particles, relative to the larger 25 and 88 nm counterparts was due to the greater number, in the smallest particles, of distorted octahedral sites in which Cu^{2+} ions are better stabilised [69]. This hypothesis correlates with the observations of the association of Hg with Fe-rich sites in nano-sized CFA, but not in larger particles [29]. Such unique material properties occurring at the nano-scale properties likely account for the metal-nanoparticle interactions observed in various investigations cited above.

Having examined the relationship between metal ions and nano-sized fractions of CFA and shown that these particles bind larger fractions of metals than micron-sized particles, let us now examine evidence that this phenomenon carries over when CFA is used for remediation of contaminated water, that is, that the nano-sized fraction is still responsible for a greater fraction of the metal removal observed. Synchrotron-based X-ray fluorescence was used to study the adsorption of Hg from flue gas by fly ash from a Kentucky utility [29]. The bulk ash (20 nm–10 μm in diameter) was largely amorphous with only quartz and hematite as crystalline phases. The fine ash had an average particle size of < 100 nm and comprised of more crystalline phases and some cementitious phases, for example, syngenite ($K_2Ca(SO_4)_2 \cdot H_2O$), glauberite ($Na_2Ca(SO_4)_2$) and picromerite ($K_2Mg(SO_4)_2$). A summary of Hg concentrations of the two classes of ash and fine ash before and after exposure to flue gas is presented in **Table 2**.

The 100 nm fraction represented only 0.01% of the total fly ash content. Nevertheless, it is clear from these data that despite already containing greater Hg concentrations before the exposure reaction, nano-sized CFA particles dominated Hg uptake from the flue gases.

Mercury was bound to hematite by bidentate inner-sphere complexes. The nature of binding is an important consideration because of its role in desorption processes as inner-sphere complexes are known to be more stable and less prone to desorption than outer-sphere complexes [43]. Using extended X-ray absorption fine structure (EXAFS) spectroscopy, Ha et al. [73] showed that Zn(II) surface complexes on 10.5 nm hematite particles were more strongly held than those adsorbed to 550 nm particles. As a result, only 10–15% of Zn^{2+} ions were desorbed from the 10.5 nm particles where they were held as inner-sphere complexes but up to 40% were lost from the larger particles where Zn^{2+} ions were held in outer-sphere complexes. This highlights the fact that nanoparticles bind metal ions differently and that this is likely to influence the long-term effectiveness of remediation activity.

	Hg concentration before flue gas exposure	Hg concentration after flue gas exposure
Bulk ash	11 ± 0.8 ppb	161 ± 4 ppb
Fine ash	100 ± 6 ppb	610 ± 20 ppb

Table 2. Hg concentrations in bulk and nano-sized particles before and after exposure to flue gases.

4. Directions for future research

As stated at the beginning of this chapter, despite the numerous studies showing the efficiency of CFA in treatment of AMD, none of those we have come across specifically describe the role of nanomaterials in this process. Nevertheless, the evidence systematically synthesised in this chapter has shown that nano-scale materials likely account for most of the remedial action of CFA. Despite the cost effectiveness and efficiency of CFA at remediation, potential secondary contamination by other must always be addressed before its application in the field. Future research should investigate the potential role of photocatalytic nanoparticles in CFA, in pollutant transformation.

Author details

Anita Etale[1*], Nikita T. Tavengwa[2] and Vusumzi E. Pakade[3]

*Address all correspondence to: aetale@gmail.com

1 ETH Zurich, Institute for Environmental Decisions (IED), Consumer Behaviour, Universitätstrasse, Zurich, Switzerland

2 Department of Chemistry, School of Mathematical and Natural Sciences, University of Venda, Thohoyandou, South Africa

3 Department of Chemistry, Vaal University of Technology, Vanderbijlpark, South Africa

References

[1] International Energy Agency, Key World Energy Statistics; 2016, 80pp

[2] Ahmaruzzaman M. A review on the utilization of fly ash. Progress in Energy and Combustion Science. 2010;**36**:327-363

[3] Blissett RS, Rowson NA. A review of the multi-component utilisation of coal fly ash. Fuel. 2012;**97**:1-23

[4] Temuujin J, Minjigmaa A, Davaabal B, Bayarzul U, Ankhtuya A, Jadambaa Ts, MacKenzie KJD. Utilization of radioactive high-calcium Mongolian fly ash for the preparation of alkali-activated geopolymers for safe use as construction materials. Ceramics International. 2014;**40**:16475-16483

[5] Durand JF. The impact of gold mining on the Witwatersrand on the rivers and karst system of Gauteng and North West Province, South Africa. Journal of African Earth Sciences. 2012;**68**:24-43

[6] Nieto JM, Sarmiento AM, Olías M, Canovas CR, Riba I, Kalman J, Delvalls TA. Acid mine drainage pollution in the Tinto and Odiel rivers (Iberian Pyrite Belt, SW Spain) and bioavailability of the transported metals to the Huelva Estuary. Environment International. 2007;**33**:445-455

[7] Singer PC, Stumm W. Acidic mine drainage: The rate-determining step. Science. 1970; **167**:1121-1123

[8] Byrne P, Wood PJ, Reid I. The impairment of river systems by metal mine contamination: A review including remediation options. Critical Reviews in Environmental Science and Technology. 2012;**42**:2017-2077

[9] Cho H, Oh D, Kim K. A study on removal characteristics of heavy metals from aqueous solution by fly ash. Journal of Hazardous Materials. 2005;**127**:187-195

[10] Gitari WM, Petrik LF, Etchebers O, Key DL, Iwuoha E, Okujeni C. Passive neutralisation of acid mine drainage by fly ash and its derivatives: A column leaching study. Fuel. 2008;**87**:1637-1650

[11] Koshy N, Singh DN. Fly ash zeolites for water treatment applications. Journal of Environmental Chemical Engineering. 2016;**4**:1460-1472

[12] Belviso C, Cavalcante F Di Gennaro S, Lettino A, Palma A, Ragone P, Fiore S. Removal of Mn from aqueous solution using fly ash and its hydrothermal synthetic zeolite. Journal of Environmental Management. 2014;**137**:16-22

[13] Visa M. Synthesis and characterization of new zeolite materials obtained from fly ash for heavy metals removal in advanced wastewater treatment. Powder Technology. 2016;**294**:338-347

[14] Attari M, Bukhari SS, Kazemian H, Rohani S. A low-cost adsorbent from coal fly ash for mercury removal from industrial wastewater. Journal of Environmental Chemical Engineering. 2017;**5**:391-399

[15] Ogata F, Iwata Y, Kawasaki N. Properties of novel adsorbent produced by hydrothermal treatment of waste fly ash in alkaline solution and its capability for adsorption of tungsten from aqueous solution. Journal of Environmental Chemical Engineering. 2015;**3**:333-338

[16] Sočo E, Kalembkiewicz J. Removal of Copper(II) and Zinc(II) ions from aqueous solution by chemical treatment of coal fly ash. Croatica Chemica Acta. 2015;**88**:267-279

[17] Pizarro J, Castillo X, Jara S, Ortiz C, Navarro P, Cid H, Rioseco H, Barros D, Belzile N. Adsorption of Cu^{2+} on coal fly ash modified with functionalized mesoporous silica. Fuel. 2015;**156**:96-102

[18] Kim HJ, Joshi MK, Pant HR, Kim JH, Lee E, Kim CS. One-pot hydrothermal synthesis of multifunctional Ag/ZnO/fly ash nanocomposite. Colloids Surfaces A: Physicochemical and Engineering Aspects. 2015;**469**:256-262

[19] Ökte AN, Karamanis D. A novel photoresponsive ZnO-flyash nanocomposite for environmental and energy applications. Applied Catalysis B Environmental. 2013;**142-143**: 538-552

[20] Qi G, Lei X, Li L, Yuan C, Sun Y, Chen J. Chen J, Wang Y, Hao J. Preparation and evaluation of a mesoporous calcium-silicate material (MCSM) from coal fly ash for removal of Co(II) from wastewater. Chemical Engineering Journal. 2015;**279**:777-787

[21] Papandreou AD, Stournaras CJ, Panias D, Paspaliaris I. Adsorption of Pb(II), Zn(II) and Cr(III) on coal fly ash porous pellets. Minerals Engineering. 2011;**24**:1495-1501

[22] Ríos CA, Williams CD, Roberts CL. Removal of heavy metals from acid mine drainage (AMD) using coal fly ash, natural clinker and synthetic zeolites. Journal of Hazardous Materials. 2008;**156**:23-35

[23] Pérez-López R, Nieto JM, de Almodóvar GR. Utilization of fly ash to improve the quality of the acid mine drainage generated by oxidation of a sulphide-rich mining waste: Column experiments. Chemosphere. 2007;**67**:1637-1646

[24] Evangelou VP. Pyrite chemistry: The key for abatement of acid mine drainage. In: Geller W, Klapper H, Salomons W, editors. Acidic Min. Lakes Acid Mine Drainage, Limnol. Reclam. 1st ed. Berlin Heidelberg: Springer-Verlag; 1998. pp. 197-222

[25] Gitari WM, Petrik LF, Etchebers O, Key DL, Okujeni C. Utilization of fly ash for treatment of coal mines wastewater: Solubility controls on major inorganic contaminants. Fuel. 2008;**87**:2450-2462

[26] Banks D, Younger PL, Egil RA, Sheila RI. Mine-water chemistry: The good, the bad and the ugly. Environmental Geology. 1997;**32**:157-174

[27] Madzivire G, Petrik LF, Gitari WM, Ojumu TV, Balfour G. Application of coal fly ash to circumneutral mine waters for the removal of sulphates as gypsum and ettringite. Minerals Engineering. 2010;**23**:252-257

[28] Madzivire G, Gitari WM, Vadapalli VRK, Ojumu TV, Petrik LF. Fate of sulphate removed during the treatment of circumneutral mine water and acid mine drainage with coal fly ash: Modelling and experimental approach. Minerals Engineering. 2011;**24**:1467-1477

[29] Jew AD, Rupp EC, Geatches DL, Jung JE, Farfan G, Bahet L, Hower JC, Brown Jr GD, Wilcox J. Mercury interaction with the fine fraction of coal-combustion fly ash in a simulated coal power plant flue gas stream. Energy and Fuels. 2015;**29**:6025-6038

[30] Lighty JS, Veranth JM, Sarofim AF. Combustion aerosols: Factors governing their size and composition and implications to human health. Journal of the Air & Waste Management Association. 2000;**50**:1565-1618

[31] Chen Y, Shah N, Huggins FE, Huffman GP, Dozier A. Characterization of ultrafine coal fly ash particles by energy-filtered TEM. Journal of Microscopy. 2005;**217**:225-234

[32] Ribeiro J, DaBoit, K Flores D, Kronbauer MA, Silva LFO. Extensive FE-SEM/EDS, HR-TEM/EDS and ToF-SIMS studies of micron- to nano-particles in anthracite fly ash. Science of the Total Environment. 2013;**452-453**:98-107

[33] Silva LFO, Oliveira MLS, Neace ER, O'Keefe JMK, Henke KR, Hower JC. Nanominerals and ultrafine particles in sublimates from the Ruth Mullins coal fire, Perry County, Eastern Kentucky, USA. International Journal of Coal Geology. 2011;**85**:237-245

[34] Silva LFO, Moreno T, Querol X. An introductory TEM study of Fe-nanominerals within coal fly ash. Science of the Total Environment. 2009;**407**:4972-4974

[35] Saikia J, Narzary B, Roy S, Bordoloi M, Saikia P, Saikia BK. Nanominerals, fullerene aggregates, and hazardous elements in coal and coal combustion-generated aerosols: An environmental and toxicological assessment. Chemosphere. 2016;**164**:84-91

[36] Ribeiro J, Flores D, Ward CR, Silva LFO. Identification of nanominerals and nanoparticles in burning coal waste piles from Portugal. Science of the Total Environment. 2010;**408**:6032-6041

[37] Bhanarkar AD, Gavane AG, Tajne DS, Tamhane SM, Nema P. Composition and size distribution of particules emissions from a coal-fired power plant in India. Fuel. 2008;**87**: 2095-2101

[38] Reddy MS, Basha S, Joshi HV, Jha B. Evaluation of the emission characteristics of trace metals from coal and fuel oil fired power plants and their fate during combustion. Jurnal of Hazardous Materials. 2005;**123**:242-249

[39] Yi H, Hao J, Duan L, Tang X, Ning P, Li X. Fine particle and trace element emissions from an anthracite coal-fired power plant equipped with a bag-house in China. Fuel. 2008;**87**:2050-2057

[40] Gilbert B. Nanoparticles: Strained and stiff. Science. 2004;**305**:651-654

[41] Banfield JF, Zhang H. Nanoparticles in the environment. Reviews in Mineralogy and Geochemistry. 2001;**44**:1-51. DOI: 10.2138/rmg.2001.44.01

[42] Waychunas GA, Zhang H. Structure, chemistry, and properties of mineral nanoparticles. Elements. 2008;**4**:381-388. DOI: 10.2113/gselements.4.6.381

[43] Brown GE, Calas G. Environmental mineralogy – Understanding element behavior in ecosystems. Comptes Rendus Geoscience. 2011;**343**:90-112

[44] Hochella Jr MF, Lower SK, Maurice PA, Penn RL, Sahai N, Sparks DL, Twining BS. Nanominerals, mineral nanoparticles, and earth systems. Science. 2008;**80**:1631-1636

[45] Brown GE, Henrich VE, Casey WH, Clark DL, Eggleston C, Felmy A, Goodman DW, Grätzel M, Maciel G, McCarthy MI, Nealson KH, Sverjensky DA, Toney MF, Zachara JM. Metal oxide surfaces and their interactions with aqueous solutions and microbial organisms. Chemical Reviews. 1999;**99**:77-174

[46] Dwivedi S, Saquib Q, Al-Khedhairy AA, Ali A-YS, Musarrat J. Characterization of coal fly ash nanoparticles and induced oxidative DNA damage in human peripheral blood mononuclear cells. Science of the Total Environment. 2012;**437**:331-338

[47] Wilcox J, Wang B, Rupp E, Taggart R, Hsu-Kim H, Oliveira MLS, Cutruneo CMNL, Taffarel S, Silva LFO, Hopps SD, Thomas GA, Hower JC. Observations and assessment of fly ashes from high-sulfur bituminous coals and blends of high-sulfur bituminous and subbituminous coals: Environmental processes recorded at the macro- and nanometer scale. Energy and Fuels. 2015;**29**:7168-7177

[48] Chen Y, Shah N, Huggins FE, Huffman GP. Transmission electron microscopy investigation of ultrafine coal fly ash particles. Environmental Science & Technology. 2005;**39**: 1144-1151

[49] Niu Y, Wang S, Shaddix CR, Hui S. Kinetic modeling of the formation and growth of inorganic nano-particles during pulverized coal char combustion in O_2/N_2 and O_2/CO_2 atmospheres. Combustion and Flame. 2016;**173**:195-207

[50] Linak WP, Wendt JOL. Toxic metal emissions from incineration: Mechanisms and control. Progress in Energy and Combustion Science. 1993;**19**:145-185. DOI: 10.1016/0360-1285(93)90014-6

[51] Nelson PF. Trace metal emissions in fine particles from coal combustion. Energy and Fuels. 2007;**21**:477-484

[52] Davison RL, Natusch DFS, Wallace JR, Evans CA. Trace elements in fly ash. Dependence of concentration on particle size. Environmental Science & Technology. 1974;**8**:1107-1113

[53] Galbreath KC, Toman DL, Zygarlicke CJ, Pavlish JH. Trace element partitioning and transformations during combustion of bituminous and subbituminous U.S. coals in a 7-kW combustion system. Energy and Fuels. 2000;**14**:1265-1279

[54] Markowski GR, Filby R. Trace element concentration as a function of particle size in fly ash from a pulverized coal utility boiler. Environmental Science & Technology. 1985; **19**:796-804

[55] Fix G, Seames W, Mann M, Benson S, Miller D. The effect of combustion temperature on coal ash fine-fragmentation mode formation mechanisms. Fuel. 2013;**113**:140-147

[56] Chen Y, Wang G, Sheng C. Comparison of particle size evolution during pulverized coal combustion in O_2/CO_2 and O_2/N_2 atmospheres. Energy and Fuels. 2014;**28**:136-145

[57] Fix G, Seames WS, Mann MD, Benson SA, Miller DJ. The effect of oxygen-to-fuel stoichiometry on coal ash fine-fragmentation mode formation mechanisms. Fuel Processing Technology. 2011;**92**:793-800

[58] Kazanc F, Levendis YA. Physical properties of particulate matter emitted from combustion of coals of various ranks in O_2/N_2 and O_2/CO_2 environments. Energy and Fuels. 2012;**26**:7127-7139

[59] Neville M, Quann RJ, Haynes BS, Sarofim F. Vaporization and condensation of mineral matter during pulversized coal combustion. Symposium (International) on Combustion. 1981;**18**:1267-1274

[60] Kazanc F, Levendis YA, Maffei T. Chemical composition of submicrometer particulate matter (PM_1) emitted from combustion of coals of various ranks in O_2/N_2 and O_2/CO_2 environments. Energy and Fuels. 2013;**27**:4984-4998

[61] Wen C, Fan B, Wang W, Zeng X, Yu G, Lv W, Minghou X. Preliminary research on the effects of coal devolatilization and char combustion processes on the emission of particulate matter during lignite combustion under air and oxy-fuel conditions. Energy and Fuels. 2017;**31**:24-230

[62] Wen C, Xu M, Yu D, Sheng C, Wu H, Zhang P, et al. PM_{10} formation during the combustion of N_2-char and CO_2-char of Chinese coals. Proceedings of the Combustion Institute. 2013;**34**:2383-2392

[63] Oliveira MLS, Marostega F, Taffarel SR, Saikia BK, Waanders FB, DaBoit K, Baruach BP, Silva LFO. Nano-mineralogical investigation of coal and fly ashes from coal-based captive power plant (India): An introduction of occupational health hazards. Science of the Total Environment. 2014;**468-469**:1128-1137

[64] Martinello K, Oliveira MLS, Molossi FA, Ramos CG, Teixeira EC, Kautzmann RM. Silver LFO. Direct identification of hazardous elements in ultra-fine and nanominerals from coal fly ash produced during diesel co-firing. Science of the Total Environment. 2014;**470-471**:444-452

[65] Saikia BK, Ward CR, Oliveira MLS, Hower JC, Baruah BP, Braga M, Silver LF Geochemistry and nano-mineralogy of two medium-sulfur northeast Indian coals. International Journal of Coal Geology. 2014;**121**:26-34

[66] Linak WP, Yoo JI, Wasson SJ, Zhu W, Wendt JOL, Huggins FE, Chen Y, Shah N, Huffman GP, Gilmour MI. Ultrafine ash aerosols from coal combustion: Characterization and health effects. Proceedings of the Combustion Institute. 2007;**31**(2):1929-1937

[67] Bhangare RC, Ajmal PY, Sahu SK, Pandit GG, Puranik VD. Distribution of trace elements in coal and combustion residues from five thermal power plants in India. International Journal of Coal Geology. 2011;**86**:349-356

[68] Madden A, Hochella M. A test of geochemical reactivity as a function of mineral size: Manganese oxidation promoted by hematite nanoparticles. Geochimica et Cosmochimica Acta. 2005;**69**:389-398

[69] Madden AS, Hochella MF, Luxton TP. Insights for size-dependent reactivity of hematite nanomineral surfaces through Cu^{2+} sorption. Geochimica et Cosmochimica Acta. 2006;**70**:4095-4104

[70] Waychunas GA, Kim CS, Banfield JF. Nanoparticulate iron oxide minerals in soils and sediments: Unique properties and contaminant scavenging mechanisms. Journal of Nanoparticle Research. 2005;**7**:409-433

[71] Weaver RM, Hochella MF, Ilton ES. Dynamic processes occurring at the Cr(III)aq-manganite (ɣ-MnOOH) interface: Simultaneous adsorption, microprecipitation, oxidation/reduction, and dissolution. Geochimica et Cosmochimica Acta. 2002;**66**:4119-4132

[72] Junta JL, Hochella Jr MF. Manganese(II) oxidation at mineral surfaces: A microscopic and spectroscopic study. Geochimica et Cosmochimica Acta. 1994;**58**:4985-4999

[73] Ha J, Trainor TP, Farges, F, Brown GE. Interaction of Zn (II) with hematite nanoparticles and microparticles: Part 2. ATR-FTIR and EXAFS study of the aqueous Zn(II)/oxalate/hematite ternary system. Langmuir. 2009;**25**:5586-5593

Chapter 3

Chemical Stabilization of Coal Fly Ash for Simultaneous Suppressing of As, B, and Se Leaching

Sri Hartuti, Shinji Kambara, Akihiro Takeyama,
Farrah Fadhillah Hanum and
Erda Rahmilaila Desfitri

Additional information is available at the end of the chapter

http://dx.doi.org/10.5772/intechopen.72135

Abstract

The discard of coal fly ash produced from the combustion of pulverized coal in a coal-fired boiler of thermal power plants has led to environmental concerns. Due to the interaction of fly ash particles with weathering and hydrological processes, the rainfall leaches out toxic elements in coal fly ash from the ash heaps. This situation has been pointed out as a potential contamination of soil, surface, and groundwater. In this chapter, the available fly ash treatment techniques to minimize future release of toxic trace elements (arsenic, boron, and selenium) have been documented, and the recent investigations dealing with leaching suppression effect of arsenic, boron, and selenium from coal fly ash have been reviewed. The leaching characteristics of arsenic, boron, and selenium are discussed, and a simple and low-cost leaching control method is presented in the context of treating the fly ash through chemical stabilization technique using additives containing high levels of calcium. Experimental results described in this chapter show the chemical stabilization technique utilizing Ca-containing additives is an effective technique for simultaneous suppressing of As, B, and Se leaching from coal fly ash.

Keywords: chemical stabilization technique, coal fly ash, leaching, calcium, arsenic, boron, selenium

1. Introduction

Coal fly ash from coal burning power generation is one of the major sources of environmental pollution due to the discharge of large amounts of fly ash into the environment. Coal fly ash has been utilized in different ways such as a substitute material for Portland cement, structural fills (usually for road construction), soil stabilization, and mineral filler in asphaltic concrete because of its physical (selfhardening) properties [1, 2], and in recent years, coal fly ash has been utilized as a potential material to treat acid mine drainage (AMD) because of its chemical (high degree of alkalinity) properties to neutralize AMD [3–8]. However, most of the fly ash is disposed in landfills [1], this disposal involves the interaction of fly ash particles with weathering and hydrological processes where rainfall leaches out toxic elements, anions, and cations from the ash heaps, which pose an environmental hazard through contamination of soil, surface, and groundwater [9–11]. During combustion, the organic matter in coal is destroyed, and as a result, the concentrations of trace elements are increased relative to the source coal. Among the elements that can be leached from fly ash, Ag, As, B, Ba, Cd, Co, Cr, Cu, Hg, Ni, Pb, Se, Sb, Sn, and Zn are of the greatest concern [12, 13] as environmental hazard. The leaching of trace elements such as As, B, and Se from coal fly ash (CFA) is likely to occur as these elements tend to form hydrophilic oxides that are dissolved as oxyanion forms [10]. The beneficial reuse of fly ash as a potential material to treat acid mine drainage (AMD) has great potential in minimizing the amount of disposed fly ash [3–8]; however, since the coal fly ash itself contain leachable trace elements and the usage of fly ash in treating AMD could lead to trace element accumulation with negative consequences to the environment [14], it is particularly important to be able to assess the leachability of coal fly ash (since the results can determine not only whether the ash is environmentally acceptable for use as the soil supplements, construction material, or neutralization material to treat AMD but to extend of isolation that might be required for disposal in landfill) and to treat the coal fly ash before the utilization to avoid the trace elements leaching into the environment.

To predict the possible effect of coal fly ash on the environment, it is particularly important to understand the factors that control the leaching behavior of trace elements in coal fly ash. Research studies on the leaching behavior of As, B, and Se in coal fly ash have been carried out with the promising results. The results demonstrated that the leaching behavior of arsenic, boron, and selenium from fly ash was affected by pH, solid-to-liquid ratio, leaching time, and ash type [15–19]. The leaching of As increased with increasing pH values in acidic fly ashes [19] and increased with decreasing pH values in alkaline fly ashes [20], while the leaching of B decreased with increasing pH values [18, 21] and the leaching of Se tends to decrease as pH was raised for an alkaline ash with high Ca composition [18]. The leaching of As, B, and Se from CFA generally increased with increases in the S/L ratio and leaching time, and adsorption/desorption played a major role in As and Se leaching from the CFA [22]. The leaching of As and Se from acidic ashes could be described by sorption of iron oxide, while the leaching from the alkaline ashes appeared to be controlled by sorption in the alkaline calcium phase [20, 23]. The presence of Ca in fly ash plays an important role in the leaching behavior of As, B, and Se, in which the leaching of As, B, and Se may involve the trapping of As, B, and Se species by the ettringite phase ($3CaO{\cdot}Al_2O_3{\cdot}3CaSO_4{\cdot}32H_2O$), leading to a decrease in leaching under alkaline conditions [18, 19, 24–28]. Therefore, Ca content and the sorption process are known to play important roles in the release of As, B, and Se from CFA.

Different treatments and stabilization processes of fly ash have been proposed by the scientific community. The most common used techniques for removing toxic elements from fly ash and APC (air pollution control) residue are (1) extraction and separation, (2) chemical stabilization, (3) solidification, and (4) thermal treatment [29]. Several studies have chemically treated fly ash produced in power plants to immobilize the toxic trace elements before disposal. These studies used chemicals such as calcium aluminates [30], phosphoric acid [31], NaOH [32], and chelating surfactants such as ED3A [33]. Arsenic trioxide contaminated soil was successfully treated with a ferrous sulfate solution to promote the formation of insoluble metal-bearing phases [34] and the coal fly ash treatment with ferrous sulfate shown to be an effective method for the sequestration of As, B, Cr, Mo, Se, and V associated with coal fly ash, where the mobility of As, B, Cr, Mo, Se, and V were substantially reduced [35]. Recently, the coal fly ash chemical treatment using other waste sources containing high level of calcium has been applied [27]. The advantage of this technique is that it can be used for reducing As, B, and Se leaching at low cost with the abundant chemical available, since it utilizes wastes from other industry. The aims of this chapter are to review what is known about the factors which control the leaching of As, B, and Se in fly ashes and the method to reduce As, B, and Se leaching from fly ash. Experimental results of simultaneously leaching suppression of As, B, and Se were described in this chapter.

2. Coal fly ash and overview of leaching characteristics of arsenic, boron, and selenium from coal fly ash

2.1. Coal fly ash

Coal fly ash, a by-product of coal-fired power plants, produced from the combustion of pulverized coal in a coal-fired boiler of thermal power plants. The fly ash particles are removed from the flue gases using electrostatic precipitators, FGD systems, or bag houses and are collected and stored dry for recycling. Fly ash consists of fine particles, predominantly spherical in shape, either solid or hollow, ranging in diameter from <1 μm up to 150 μm formed from the mineral matter in coal, consisting of the noncombustible matter in coal plus a small amount of carbon that remains from incomplete combustion. Properties of fly ash vary significantly with coal composition and plant-operating conditions. Fly ash contains the primary inorganic components of SiO_2, Al_2O_3, Fe_2O_3, and CaO, less amount of MgO, Na_2O, K_2O, SO_3, MnO, TiO_2, and C and varying levels of trace elements [36–38]. Based on its chemical composition (ASTM C618), fly ash can be classified into two classes, C and F. Class C ash (high-calcium, ≥10% CaO) is normally produced from lignite or subbituminous coals and contains less $SiO_2 + Al_2O_3 + Fe_2O_3$ (>50%) but more calcium hydroxide or lime and higher amount of alkalis (combined sodium and potassium). Class F ash (low-calcium, <10% CaO) is generally produced from burning anthracite or bituminous coal and contains at least 70% of $SiO_2 + Al_2O_3 + Fe_2O_3$. [38–43]. The mineralogy of fly ash is greatly influenced by the parent coal from which it was derived. Owing to the rapid cooling of burned coal in the power plant, fly ashes consist of amorphous glass (≤ 90%) and a small amount of crystalline material [44–49]. This predominant portion of the glass gives fly ash its pozzolanic properties (harden with water after activation with an alkaline substance such as lime) [50]. The major crystalline phases in fly ashes are quartz (SiO_2), mullite ($3Al_2O_3{\cdot}2SiO_2$), magnetic spinel includes; magnetite (Fe_3O_4), and hematite (Fe_2O_3). Although the total percentage of magnetic matrix of the ash is small, particular attention should be given because of its reactivity and potential for carrying and releasing toxic elements. [47].

Fly ash particle surfaces are often enriched in highly environmentally reactive trace elements (such as As, Se, Cd, Cr, Ni, Sb, Pb, Sn, Zn, and Mo [10, 51]) due to the largely emission of various trace elements in coal during coal combustion [52–55]. It has been reported that trace element concentrations in fly ash are sometimes 4–10 times higher than their original concentrations in coal samples due to the condensation of elements, which are volatilized during combustion, onto solid particles at different rates and in varying amounts as the combustion gases cool down [56–60]. As a result, these elements readily react when the fly ash particle is exposed to water and release into the environment [35, 59]. However, the release of the elements from fly ashes is controlled by precipitation or dissolution mainly and possibly by desorption. Understanding the factors that control the leaching behavior of trace elements is critical in predicting the potential impacts of fly ash on the environment. Since As, B, and Se have recently become a major problem in soil contamination in Japan, the leaching behavior of As, B, and Se has been growing in the recent investigations and described in the following sections.

2.2. Overview of leaching characteristic of arsenic, selenium, and boron from coal fly ash

Several mechanisms have been observed to elucidate arsenic, boron, and selenium leaching behavior from fly ash. Generally, the leaching behavior of As, B, and Se was affected by pH, solid-to-liquid ratio, leaching time, and ash properties, in which these ash properties including chemical composition, chemical/mineralogical speciation, the particle morphology, and the fraction of a species are available for leaching [16–19, 26, 28, 61].

Arsenic and Selenium releases from acidic fly ash increase with pH, whereas in alkaline fly ash, this trend is reversed and the leaching of As and Se from acidic fly ash could be described by sorption on iron oxide, while leaching from alkaline fly ash seems to be controlled by sorption on an alkaline Ca phase [20]. This finding is relevant with that from Zielinski et al., where in a highly acidic fly ash, the mode of occurrence of arsenic is associated with iron oxide, oxyhydroxide, or sulfate, while in a highly alkaline ash, arsenic is associated with a phase similar to calcium arsenate (detected using XAFS (X-ray absorption fine structure) spectroscopy) [62]. It is well known that Ca-rich ash tends to make the formation of insoluble Ca-arsenate [63], whereas low-lime fly ash provides less chance for this phase to precipitate. At $pH > 11.5$, the precipitation of ettringite ($3CaO{\cdot}Al_2O_3{\cdot}3CaSO_4{\cdot}32H_2O$) contributes to the dramatic reduction of As in leaching solution, along with other oxyanionic species [10, 64]. It is also reported that the formation of ettringite ($3CaO{\cdot}Al_2O_3{\cdot}3CaSO_4{\cdot}32H_2O$) contributes to the stabilization of selenium in subbituminous coal ash [65, 66]. The adsorption of selenite in fly ash was also likely controlled by aluminum oxide [23, 51, 67].

Boron is the most mobile trace element in coal ash [21], since it is associated frequently with the smallest particles in the ash, where it can accumulate on the surfaces of particles and in water-soluble fraction which promote high leachability rates [51]. Although the leachability of B does not depend significantly upon pH at nearly neutral values (pH 6–8), the initial rate of leaching is increased by an increase in acidity and it decreases with further increase in pH, and the trend shows no substantial difference between acidic and alkaline ash [21]. On the contrary, the boron leachability depends on its element concentration in coal fly ash [18], and the ligand exchange mechanisms are considered as a reason for the lower concentrations of B under mildly acidic conditions [68, 69], although much greater leaching rates are achieved when under strongly acidic conditions [69, 70]. Previous studies indicate the tendency for ettringite to incorporate B within its mineral lattice and lead to the decreasing of solubility/dissolved concentrations of B at pH ≈ 11.5 [18, 71]. It has been also indicated that under alkaline conditions, the co-precipitation of B with $CaCO_3$ may significantly captured B in alkaline fly ash [18], to some extent, in acidic-natured fly ash, this process is unlikely to take place [19].

3. Arsenic, boron, and selenium: properties, contamination pathways, and their harmful effects

3.1. Arsenic

Arsenic is a semi metallic element with atomic number 33, atomic mass 74.92 g mol^{-1} and density 5.72 g cm^{-3} at 14°C [72]. It is an odorless, tasteless, and notoriously poisonous metalloid with various allotropic forms: black, yellow, and gray, in which the gray form is the most common, and it is used for industrial purposes. It is distributed throughout our environment, mainly in earth crust, air, water, soil, sediments, etc. Arsenic is emitted into the atmosphere by high-temperature processes such as coal-fired power generation plants, burning vegetation, and volcanism [73]. Some forms of arsenic are inorganic which do not contain carbon, and others are organic, which always contain carbon. The examples for organic arsenic compounds are arsanilic acid (4-aminophenylarsonic acid, $C_6H_8AsNO_3$), arsenobetaine (2-trimethylarsoniumylacetate, $C_5H_{11}AsO_2$), methylarsonic acid (monomethylarsonate, CH_5AsO_3), etc. [74]. Inorganic arsenic compound exists in four oxidation states: −3 (arsenide), 0 (metallic arsenic), +3 (arsenite) and +5 (arsenate). The examples for arsenide compounds are alkali and alkaline earth metal arsenides (e.g., sodium arsenide Na_3As), arsenides of group III elements (e.g., gallium arsenide GaAs) [75], etc. The most common forms of arsenite compounds are arsenic sulfide (As_2S_3), arsenic trichloride ($AsCl_3$), potassium arsenite ($AsO_2{\cdot}K$), sodium arsenite ($AsO_2{\cdot}Na$), etc. [76]. The general form of arsenate is AsO_4^{3-} and exists in different forms such as arsenic acid (H_3AsO_4) at strong acidic condition, dihydrogen arsenate (H_2AsO4^-) at weak acidic condition, hydrogen arsenate ($HAsO_4^{2-}$) at weak basic condition, and arsenate ion ($AsO4^{3-}$) at strong basic condition. Examples of arsenate compounds are arsenic pentoxide (As_2O_5), calcium arsenate ($(AsO_4)_2{\cdot}3Ca$), lead arsenate ($HAsO_4{\cdot}Pb$), potassium arsenate ($H_2AsO_4{\cdot}K$), sodium arsenate ($H_2AsO_4{\cdot}Na$), etc. [76]. Arsenate is thermodynamically stable at aerobic condition, while arsenite is stable at anaerobic conditions [77]. Generally, inorganic forms of arsenic are more toxic to the environment than organic forms, and among inorganic forms, arsenite is more toxic than arsenate due to higher cytotoxic, genotoxic, mobile, and soluble nature of arsenite [78].

Humans may be exposed to arsenic through the usage of arsenic containing water for drinking purposes, food, cosmetics, cigarettes, etc. [79]. The sources of arsenic contamination in water medium can be classified as natural and anthropogenic sources. The natural sources of arsenic contamination in water medium are due to the mobilization of arsenic in natural condition which depends on its oxidation states, mobilization mechanism, and the parent mineral form [80]. The anthropogenic sources of arsenic contamination in water medium include the wastes generated from industries manufacturing wool and cotton, glass, ceramics, semiconductor, pesticide, etc. and from industries like rare earth industry, chemical industries, petroleum refining, etc. [81, 82].

The health effects on arsenic may come both from different forms of arsenic (inorganic and organic forms) and categorized based on the level of contamination. Among the effect of inorganic arsenic exposure, disturbance of the stomach and digestive organs, diminished generation of red and white blood cells, skin changes, and lung irritation have been reported in many health cases. The take-up of significant amounts of inorganic arsenic has been proposed to increase the possibility of cancer development, particularly the possibility of skin cancer development, lung, liver, and lymphatic cancer. Barrenness and miscarriages with women, skin irritation, declined protection from diseases, heart disruptions, brain damage with both men and women, and DNA damage have been reported as the effects of a very high exposure to inorganic arsenic. The effects of organic arsenic are almost the same with inorganic arsenic, which can cause neither cancer nor DNA damage, but high doses exposure may cause nerve injury.

3.2. Boron

Boron is a nonmetallic element with atomic number 5, atomic mass 10.81 g mol^{-1}, density 2.3 g cm^{-3} at 20°C, and a poor electrical conductor at room temperature. It has several allotropes in the form of amorphous boron, a dark powder, unreactive to oxygen, water, acids, and alkalis; crystalline boron, silvery to black, and extremely hard. It is found in nature in a low concentration in oceans, earth crust, rock, soil, and water [83] and mostly in the form of over 200 minerals with different amounts of calcium, sodium, or magnesium and available as calcium, sodium, and magnesium borates. Among them, the most popular are borax, tincal, colemanite, ulexite, and kernite [84]. Boron has various oxidation states in compounds, but the most significant and common is +3. It appears in lower oxidation states +1, 0, or less than 0, but these states are found in compounds such as higher borates only [85, 86]. Boron enters the environment mainly through the weathering of rocks, boric acid volatilization from seawater, and volcanic and geothermal activity. Boron is released to the environment from anthropogenic sources e.g., via industrial air emissions, fertilizer and herbicide applications, and municipal and industrial wastes from industries manufacturing borosilicate glass, detergents, semiconductor, cosmetics, flame retardants, fertilizers, and dyestuff production [87]. Two anthropogenic boron compounds, boron trichloride and boron trifluoride, are listed as toxics release inventory (TRI) chemicals.

Human causes of boron contamination include releases to air from power plants, chemical plants, and manufacturing facilities. Contamination of water can come directly from industrial wastewater and municipal sewage, as well as indirectly from air deposition and soil runoff. Borates in detergents, soaps, and personal care products can also contribute to the presence of boron in water. Boron can be found in surface water in the form of undissociated orthoboric acid, partially dissociated borate anions in the form of polyborates, complexes of transition metals, and fluoroborate complexes, and it is also found in ground water, brackish water, or hot springs, especially at geothermal or tectonic areas [88].

Among the effect of long-term consuming of food and water with increased boron content, cardiovascular, coronary, nervous, reproductive systems, changes in blood composition, neurological effects, physical disorders, and intellectual development of children have been reported in many health problems. For pregnant women, abundance of boron can be notably harmful as it raises the risk of birth pathology. Testicular decay and degeneration have been detected as the effects of high daily doses of boron. Queasiness, retching, diarrhea, and lack of appropriate weight gain or weigh loss have been detected as the effects of consuming of food and water with doses greater than 500 mg/day [89].

3.3. Selenium

Selenium is a nonmetallic element with atomic number 34, atomic mass 78.96 g mol^{-1}, and density 4.79 g cm^{-3} at 20°C. Selenium has a number of allotropes including a gray crystalline hexagonal selenium, a red crystalline form, an amorphous (which has a disordered atomic structure) red powder, and a black vitreous (glass-like) form. Amorphous selenium is a photoconductor (a light-dependent semiconductor), and it has had a long history of use in light-based applications—it was used to make the first solar cell in 1883. Selenium is a rare component of the Earth's crust that is mostly found as selenide associated with heavy metal sulfide ores, such as copper and lead sulfides [90]. Selenium species can be grouped into the four major categories: (1) inorganic selenium, (2) volatile and methylated selenium, (3) protein and amino acid selenium, and (4) nonprotein amino acids and biochemical intermediates. Selenium compounds commonly exist in the oxidation states −2, +2, +4, and +6. It is usually found as the oxyanions selenate (SO_4^{2-}) and selenite (SO_3^{2-}) in oxidized systems and as elemental selenium (Se[0]) and selenides (HSe-) in aerobic zones and unweathered mineral formations [91]. Though complexed selenium is of low toxicity, selenate (SeVI) and selenites (SeIV) are very toxic. These two forms of selenium are generally found in water and display bioaccumulation and bioavailability. Under acidic conditions, the extremely toxic and corrosive hydrogen selenide gas can be generated from selenium containing species.

Selenium contamination typically occurs in the aqueous stream and bioaccumulation of selenates and selenites in waste water can threat all aquatic life downstream. Natural processes that redistribute selenium include volcanic activity, terrestrial weathering of rocks and soils, wildfires, and volatilization from plants and water bodies. The anthropogenic sources of selenium to aquatic systems are including mining, fossil fuel combustion, oil refining, and discharge of seleniferous drainage water from irrigated agriculture [92]; these sources will end up in groundwater or surface water through irrigation.

Humans may be exposed to selenium through food or water or contact with soil or air that contains high concentrations of selenium. The health effects of various forms of selenium can vary from brittle hair and deformed nails, to rashes, heat, swelling of the skin, and severe pains. Exposure to selenium through air can cause dizziness, fatigue, and irritations of the mucous membranes. Among the effect of selenium overexposure, red staining of the nails, teeth, and hair have been reported in many health cases. Due to the fact that selenium is an eye and upper respiratory irritant and a sensitizer, selenious acid produced from the reaction of selenium dioxide with moisture may cause corrosive to the skin and eyes. Accumulation of fluid in the lungs, pneumonitis, bronchitis, bronchial asthma, shortness of breath, sore throat, chills, fever, headache, conjunctivitis, queasiness, retching, abdominal pain, diarrhea, and enlarged liver has been reported as the effects of overexposure of selenium fumes.

4. Fly ash treatment techniques

Treatment of fly ash before utilization is important to minimize leaching of contaminants and/or utilize in the best possible manner in which methodologies for detoxifying fly ash including reducing the concentrations of the contaminants (e.g., through washing), astriction the leaching of the contaminants by forming steady blocks or inertial compounds with additives or binders (e.g., by stabilization methods), or reducing the mobility of the contaminants (e.g., through S/S methods) [93]. Several researchers have produced numerous treatments and disposal solutions for fly ash over the recent decades, some of these are only tested in the laboratory, while others are available commercially. Based on the main principle of operation, the treatment techniques may be classified in four categories [29, 94]:

1. extraction and separation,
2. chemical stabilization,
3. solidification, and
4. thermal treatment.

4.1. Extraction and separation

Extraction and separation techniques have the main purpose to remove or recover specific components or fractions from fly ash which focus on removing heavy metals and to some extent salts from fly ash, mainly using water or acidic solutions. The processes of this treatment technique are typically relatively simple, but the main disadvantage is the generation of process water with high content of metals and salts.

The main extracting agent used in this technique including water [95], acid [96], and microorganisms [97]. The electrodialysis process which involve ion exchange technique [98] and particle size fractionation process based on the settling velocities of fly ash particles in a water filled reactor [99] are also including in this extraction and separation techniques. For the comprehensive overview of the principles of these techniques, see Refs. [94, 100].

4.2. Chemical stabilization

Chemical stabilization technique has the main purpose to bind and immobilize pollutants in the fly ash matrix which focus on binding heavy metals by alterations to the fly ashes geochemical properties [101, 102]. Generally, the processes of the chemical stabilization technique are simple and low cost, which significantly improve the leaching properties of the fly ash, such as water extraction, chemical reactions, and then, de-watering. The main disadvantage is the generation of metal and salt containing process water.

Various chemical stabilization processes have been developed, and most of them involve the use of $FeSO_4$ [35, 95, 103], $FeSO_4 + CaCO_3$ [35], CO_2 [104, 105], CO_2, and H_3PO_4 [106] Phosphate (PO_4^{3-}) [107], Sulfide (S^{2-}) [108], $Ca(OH)_2$, and paper sludge ashes [27], $Ca(OH)_2$ and cement [109, 110], colloidal silica [111], and silica fume [112].

4.3. Solidification

Solidification technique is often discussed as stabilization processes (S/S: solidification and stabilization) in which solidification involves the transformation of a liquid or sludge into solids and may not lead to a chemical interaction of the constituent of concern with the solidifying agent. This technique has the main purpose to physically and hydraulically encapsulate the fly ashes and reduces the mobility of the contaminants in fly ashes because of the reduced surface area and low permeability. On the other hand, the main goal of stabilization is to convert the contaminants into less soluble or less toxic forms, with or without solidification [113]. The main advantages of this technique come from the fact that this technology is simple, well established, and low cost, which significantly decrease the leaching of contaminants and improve the mechanical properties. The main disadvantages of this technique are related to significant increase in the mass disposed of, and the physical integrity of the product may deteriorate over time resulting in increased metals leaching.

The main solidifying agent used in this technique including water [114], cement [93], and cement [115]. For the comprehensive overview of the principles of these techniques, see Refs. [94, 100, 116].

4.4. Thermal treatment

Thermal treatment technique involves a heating of fly ashes and thereby changing the physical and chemical properties of fly ashes in which the stable and very dense product can be produced with sufficient leaching properties [117]. This method very efficient at destroying dioxins, furans, and other toxic organic compounds due to the high temperatures used [118]. In some cases, encapsulation of fly ash is also occurred [101]. The main disadvantages are the high cost due to the high-energy demands for the process and generation of flue gas containing volatile metals.

Three major types of thermal treatment technique are sintering, vitrification, and melting.

1. Sintering processes involves heating to a level at which individual particles are bound together and the chemical species of interest can achieve a reconfiguration [119]. Temperatures are around 900–1300°C, and a denser and less porous material is produced.

2. Vitrification processes involves melting of a mixture of fly ash and additives (glass precursors (i.e., Si)) to fix the contaminants in the final matrix (alumina silicates). This mixture is typically heated to around 1300–2000°C and then cooled to form a single solid phase (an amorphous and homogenous glassy material). In this process, fly ash components are bound in the glassy materials thereby also encapsulating the fly ash.

3. Melting (or fusion) processes are very similar to vitrification processes; however, in this case, no additives (glass forming materials) are added. The final product consists of multiple metal phases [118] which can possibly be separated utilizing different melting temperature of individual metal phases. Therefore, the product obtained can be used as a construction material, for example, as a sub-base in road construction or for pavements [120]. The temperatures involve in this melting processes are like vitrification processes.

5. Experimental results of simultaneous leaching suppression of As, B and Se from coal fly ash

Considering the leaching of arsenic, boron, and selenium from fly ash has recently become a major problem in soil contamination in Japan [27, 121], the current research focuses on the development of effective technologies for the improving leaching properties of fly ash. As described in section 4, several pre-treatment options have been proposed for minimizing the leaching of toxic trace elements, including addition of $Ca(OH)_2$ to fly ash [109, 110]. Based on this method, the Laboratory of Environmental Engineering Systems of Gifu University-Japan, one of the founding members of the Next Generation Energy Research Center, developed its scientific research on "Experimental study on simultaneous leaching suppression of trace elements including As, B, and Se from coal fly ash" which analyzes the leaching properties and promotes the methods for control of the trace elements leachates from coal fly ash, utilizing the chemical stabilization technique by using $Ca(OH)_2$, paper sludge ash, and filter cake, which are waste generated in the papermaking and lime industry processes [27].

In this study, a low calcium content fly ash sample named fly ash H (FAH) (2.05% of CaO, detected using X-ray fluorescence) with a high concentration of trace elements leaching (As 48.66 μg/L, B 5.39 mg/L, Se 86.9 μg/L, detected using ICP-AES) that was collected from coal-fired power plant in Japan was treated through chemical stabilization technique with the addition of additives ($Ca(OH)_2$, three kinds of paper sludge ashes (PS Ash 3, 4, 5) and one kind of filter cake (FC)) under 5 and 10% Ca content in additives and subjected in a detailed characterization and leaching test regarding its chemical properties and its leaching behavior. **Table 1** lists the composition of inorganic elements in FAH and additives. **Figure 1** depicts experimental process of chemical stabilization in this study. Fly ash sample and additives characterization

Sample			**FAH (1s)** [1]	**PS Ash 3**	**PS Ash 4**	**PS Ash 5**	**FC**
Ash composition	SiO_2	(%)	59.25	31.47	44.21	42.36	23.31
	Al_2O_3		25.63	12.40	22.23	19.80	13.87
	TiO_2		1.99	0.38	2.56	2.11	0.06
	Fe_2O_3		7.49	5.13	2.63	5.56	2.33
	CaO		2.05	46.31	18.77	19.51	59.18
	MgO		0.79	3.28	3.42	3.30	0.96
	Na_2O		0.60	0.24	0.95	0.41	0.03
	K_2O		1.56	0.20	2.09	1.99	0.25
	P_2O_5		0.18	0.18	1.75	1.54	0.00
	MnO		—	0.03	0.05	0.06	0.04
	V_2O_5		0.03	0.01	0.02	0.02	0.02
	SO_3		0.42	0.36	1.31	3.33	0.00
pH of the leachate			10.38	13.72	12.71	12.26	7.28
Leached Ca ion		(mg/L)	121	1405	246.5	597	15.45
Leaching concentration	As	(μg/L)	48.66				
	B	(mg/L)	5.39				
	Se	(μg/L)	86.9				

[1]Sample fly ash H, from the chamber 1 of electrostatic precipitator.

Table 1. Composition of inorganic elements in fly ash and additives.

include the following measurements: pH, elemental composition, trace elements (As, B and Se) leaching and Ca ion concentrations (**Table 1**), thermogravimetric (TG) analysis, X-ray diffraction (XRD) analysis, and leachate analysis (pH, trace elements (As, B and Se) leaching and ions concentrations). This treatment technique, using additives addition, was examined in order to evaluate their effectiveness in the improvement of fly ash leaching properties. Additives may reduce the leachability of As, B and Se contained in ash by promoting the formation of ettringite or precipitation with calcium in ash, which are resistant to leaching. For the classification of untreated and treated ash samples, the procedure of standard leaching tests for fly ash (Notification No. 13 by the Environmental Agency of Japan) was employed as the protocol for leaching tests in this study.

Figure 2 depicts the leaching concentration of As, B, and Se for fly ash H alone and fly ash H under five kinds of additives addition for 5% and 10% Ca content samples, and pH values of leachate. As indicated in **Figure 2**, the results showed that the leaching concentrations of As, B, and Se in FAH tremendously decreased below the soil environmental standard in Japan for both 5% and 10% Ca content in additives and further decreased with the increase of Ca content. Especially, the leaching concentration of As, B, and Se reduced to 91–100% with $Ca(OH)_2$ addition. Among three kinds of paper sludge ashes (PS Ash

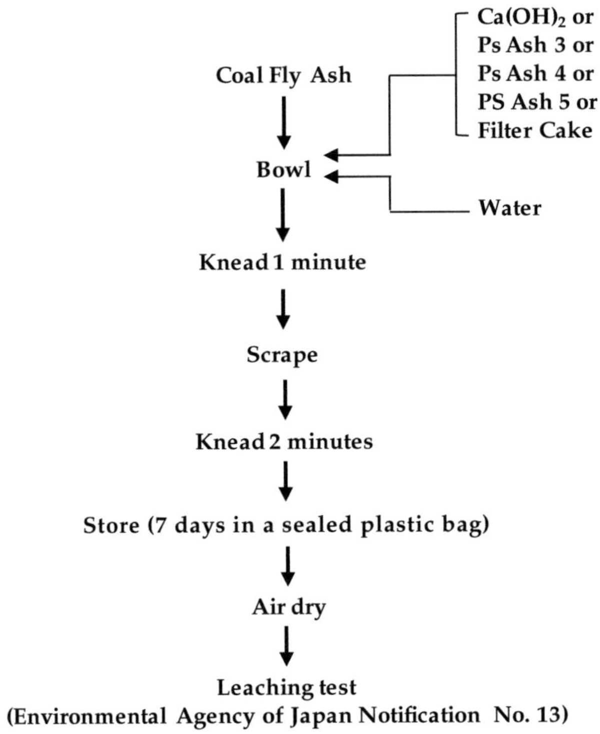

Figure 1. The experimental scheme of chemical stabilization process.

3, 4, 5) addition, As and B leaching concentrations reduced to 89–96% and 83–92%, respectively, with PS Ash 3 addition; and Se leaching concentration reduced to 87–96% with PS Ash 5 addition. While As and B leaching concentration did not influenced by filter cake (FC) addition (owing to the composition of FC is calcium carbonate based, which is not a water-soluble calcium compound (**Figure 3**)), but Se leaching concentration reduced to 58–78% (owing to the presence of Al_2O_3 and Fe_2O_3 in FC contributing for selenium to being hard to leach). **Figure 3** depicts XRD patterns of several calcium compounds and additives. According to the leaching mechanism of selenium, Al_2O_3 and Fe_2O_3 may provide additional surface area for positively charged ions to be attached through the sorption reaction process, which made contributions to lower concentration of Se in aqueous solutions (FAH-FC mixture leachates) [122–124]. It is also indicated in **Figure 2** that pH values of chemical treated products range from 9.23 to 12.98. It could be seen that the pH of mixture leachates increased with the addition of all additives excepting FC. Especially, pH values of FAH-$Ca(OH)_2$ and FAH-Ps Ash 3 mixture leachates increased to approximately 12.98 and 11.94, respectively. This exhibited that relatively high CaO content included in PS Ash 3 and $Ca(OH)_2$ itself being a water-soluble calcium compound could promote the pH values of leachates. However, FC which contained the highest CaO content did not have any influence on the pH values of mixture leachates, this was likely to be caused as calcium compound included in FC is composed of $CaCO_3$ being an insoluble substance in water (**Figure 3**). **Figure 2** demonstrated that the minimum solubility of As, B, and Se was obtained when the pH of mixture leachates became 11.5 or higher. Obviously, higher pH

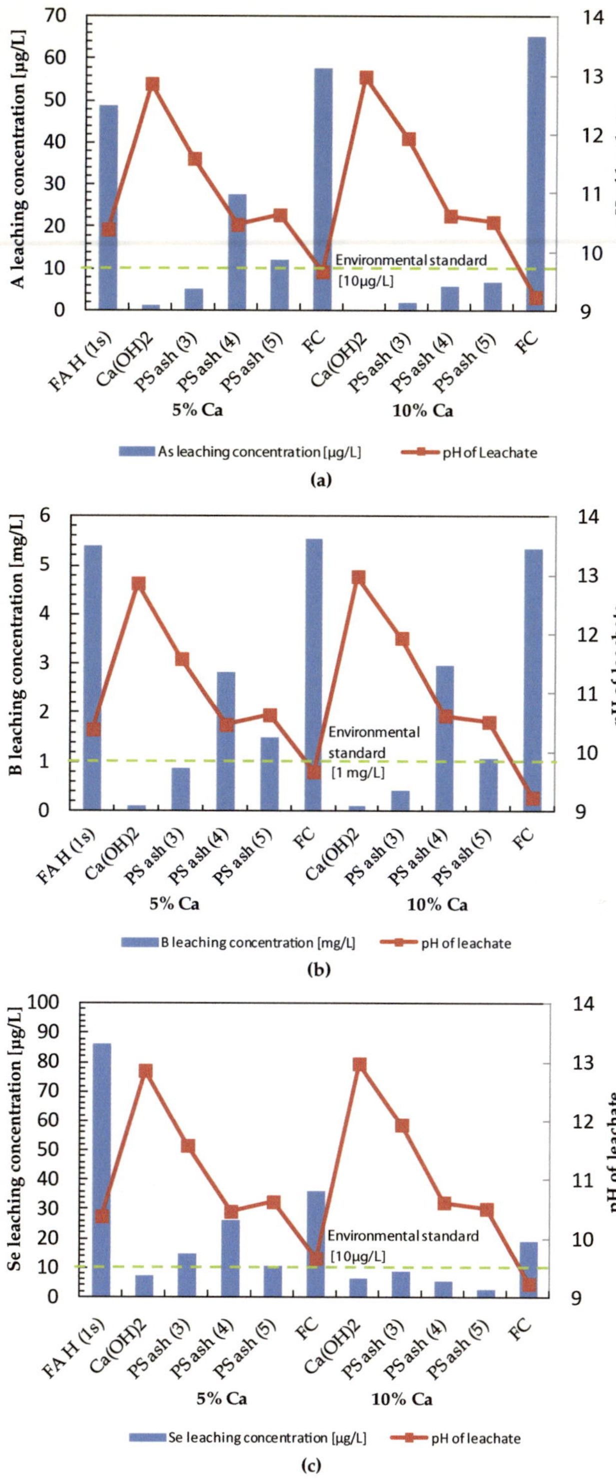

Figure 2. Trace element leaching concentration for fly ash H alone, and fly ash H under five kinds of additives addition for 5% and 10% Ca content samples: (a) arsenic; (b) boron; and (c) selenium.

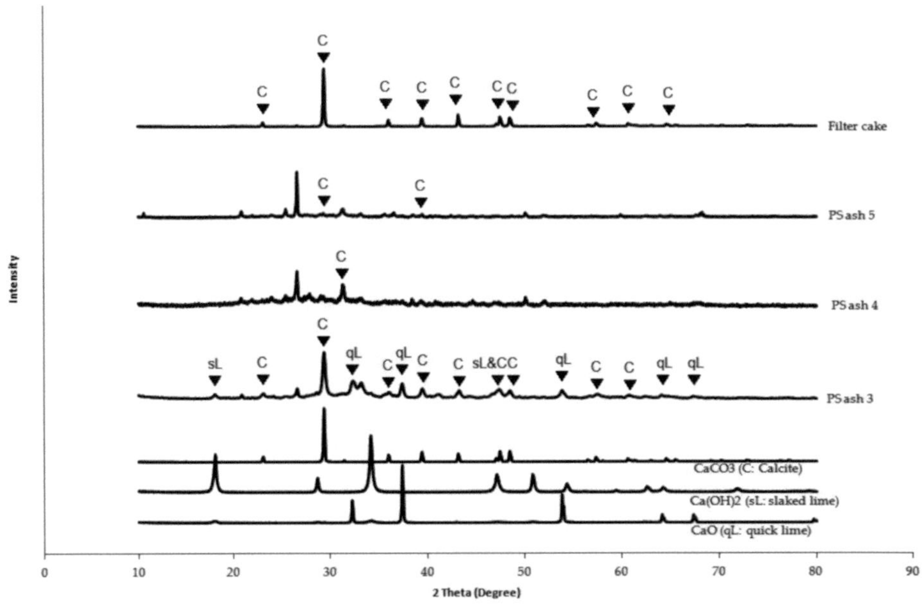

Figure 3. XRD Patterns of several calcium compounds and additives.

of mixture leachates in this study was favorable to lowering leaching concentrations of As, B, and Se and the mobility of As, B, and Se indeed decreased in **Figure 2**, implying that the efficiency of method in this study was susceptible to pH value of leachate. The results revealed that all additives (other than filter cake) showed a simultaneous leaching suppression effect of As, B, and Se from coal fly ash.

6. Conclusion

Treatment of trace elements (arsenic, boron, and selenium) contained in coal fly ash is a necessity to minimize leaching of trace elements and utilized in the best possible manner. Chemical stabilization of trace elements (arsenic, boron, and selenium) contained in coal fly ash offers a simple and low-cost environmentally friendly technique that if properly and thoroughly carried can bring our environment into a better place for both human and animal well-being due to its enormous advantages over other treatment methods. According to fly ash treatment through the addition of Ca-containing-additives experimental results, the leaching concentration of As, B, and Se can greatly be reduced simultaneously until 89–96%, 83–92%, and 87–96%, respectively. Paper sludge ash 3 as a waste generated in the papermaking process has been found as an effective and best additive (suppressing material) to reduce the leaching concentration of As, B, and Se from coal fly ash. Therefore, the chemical stabilization technique utilizing Ca-containing-additives is an effective technique for simultaneous suppressing As, B, and Se leaching from coal fly ash.

Author details

Sri Hartuti*, Shinji Kambara, Akihiro Takeyama, Farrah Fadhillah Hanum and Erda Rahmilaila Desfitri

*Address all correspondence to: amy_srihartuti@yahoo.co.id

Department of Environmental and Renewable Energy Systems, Graduate school of Engineering, Gifu University, Gifu, Japan

References

[1] World-Wide Coal Combustion Products Network (WWCCPN). 2017). [Accessed on March, 2017]. Available from: http://www.wwccpn.net/

[2] American Coal Ash Association (ACAA). An American Recycling Success Story: Beneficial Use of Coal Combustion Products; 2015

[3] Jia Y, Maurice C, Öhlander B. Effect of the alkaline industrial residues fly ash, green liquor dregs, and lime mud on mine tailings oxidation when used as covering material. Environmental Earth Sciences. 2014;**72**:319-334

[4] Prasad B, Mortimer RJG. Treatment of acid mine drainage using fly ash zeolite. Water, Air, and Soil Pollution. 2011;**218**:667-679

[5] Backstrom M, Sartz L. Mixing of acid rock drainage with alkaline ash leachates-fate and immobilisation of trace elements. Water, Air, and Soil Pollution. 2011;**222**:377-389

[6] Pérez-López R, Cama J, Miguel Nieto J, Ayora C, Saaltink MW. Attenuation of pyrite oxidation with a fly ash pre-barrier: Reactive transport modelling of column experiments. Applied Geochemistry. 2009;**24**:1712-1723

[7] Pérez-López R, Nieto JM, de Almodóvar GR. Utilization of fly ash to improve the quality of the acid mine drainage generated by oxidation of a sulphide-rich mining waste: Column experiments. Chemosphere. 2007;**67**:1637-1646

[8] Gitari MW, Petrik LF, Etchebers O, Key DL, Iwuoha E, Okujeni C. Treatment of acid mine drainage with fly ash: Removal of major contaminants and trace elements. Journal of Environmental Science and Health, Part A. 2006:1729-1747

[9] Carlson CL, Adriano DC. Environmental impacts of coal combustion residues. Journal of Environmental Quality. 1993;**22**:227-247

[10] Jones DR. (1995). The leaching of major and trace elements from coal ash, In: Swaine, DJ, Goodarzi, F. eds. Environmental Aspects of Trace Elements in Coal, Dordrecht, the Netherlands: Kluwer Academic Publisher, Vol. 228, pp. 221-262

[11] Kim AG, Kazonich G. Mass release of trace elements from coal combustion byproducts. In: International Ash Utilization Symposium. Lexington, KY: Center for Applied Energy Research, University of Kentucky; 1999

[12] Mehnert E, Hensel BR. Coal combustion by-products and contaminant transport in groundwater. In: Coal Combustion By-Products Associated with Coal Mining Interactive Forum. Carbondale: Southern Illinois University at Carbondale, IL; 1996

[13] U.S. Environmental Protection Agency (USEPA). 2014. Coal Combustion Residuals (CCR)—Surface Impoundments with High Hazard Potential Ratings. [Accessed March 1, 2015]. http://www.epa.gov/wastes/nonhaz/industrial/special/fossil/ccrs-fs/index.htm

[14] Karapanagioti HK, Atalay AS. Laboratory evaluation of ash materials as acid disturbed land amendments. Global NEST Journal: The International Journal. 2001;**3**:11-21

[15] Brunori C, Balzamo S, Morabito R. Comparison between different leaching/extraction tests for the evaluation of metal release from fly ASH. International Journal of Environmental Analytical Chemistry. 1999;**75**:19-31

[16] Baba A, Kaya A. Leaching characteristics of fly ash from thermal power plants of Soma and Tunçbilek, Turkey. Environmental Monitoring and Assessment. 2004;**91**:171-181

[17] Otero-Rey JR, Mato-Fernandez MJ, Moreda-Pineiro J, Alonso-Rodriguez E, Muniategui-Lorenzo S, Lopez-Mahia P, Prada-Rodriguez D. Influence of several experimental parameters on arsenic and selenium leaching from coal fly ash samples. Analytica Chimica Acta. 2005;**531**:299-305

[18] Iwashita A, Sakaguchi Y, Nakajima T, Takanashi H, Ohki A, Kambara S. Leaching characteristics of boron and selenium for various coal fly ashes. Fuel. 2005;**84**:479-485

[19] Jankowski J, Colin RW, French D, Groves S. Mobility of trace elements from selected Australian fly ashes and its potential impact on aquatic ecosystems. Fuel. 2006;**85**:243-256

[20] Van der Hoek EE, Bonouvrie PA, Comans RNJ. Sorption of As and Se on mineral components of fly ash: Relevance for leaching processes. Applied Geochemistry. 1994;**9**:403-412

[21] Cox JA, Lundquist GL, Przyjazny A, Schmulbach CD. Leaching of boron from coal ash. Environmental Science & Technology. 1978;**12**:722-723

[22] Wang T, Wang J, Tang Y, Shi H, Ladwig K. Leaching characteristics of arsenic and selenium from coal fly ash: Role of Calcium. Energy & Fuels. 2009;**23**:2959-2966

[23] Van der Hoek EE, Commans RNJ. Modelling As and Se leaching from acidic fly ash by sorption on iron (hydr) oxide in the fly ash matrix. Environmental Science & Technology. 1996;**30**:517-523

[24] Jiao F, Ninomiya Y, Zhang L, Yamada N, Sato A, Dong Z. Effect of coal blending on the leaching characteristics of arsenic in fly ash from fluidized bed coal combustion. Fuel Processing Technology. 2012;**106**:769-775

[25] Myneni SCB, Traina SJ, Waychunas GA, Logan TJ. Vibrational spectroscopy of functional group chemistry and arsenate coordination in ettringite. Geochimica et Cosmochimica Acta. 1998;**62**:3499-3514

[26] Hartuti S, Kambara S, Takeyama A, Hanum FF. Leaching characteristic of arsenic in coal fly ash. Journal of Materials Science and Engineering B. 2017a;**7**:19-26

[27] Hartuti S, Hanum FF, Takeyama A, Kambara S. Effect of additives on arsenic, boron and selenium leaching from coal fly ash. Minerals. 2017b;**7**:99

[28] Hartuti S, Takeyama A, Kambara S. Arsenic leachability of coal fly ashes from different types of coal fired power plants. Journal of Materials Science and Engineering A. 2017c;**7**: 169-177

[29] Astrup T. Management of APC Residues from W-t-E Plants - An Overview of Management Options and Treatment Methods. 2nd ed; Produced by members of ISWA-WG (International Solid Waste Association-Working Group Thermal Treatment of Waste, Subgroup on APC Residues from W-t-E plants). ISWA, Copenhagen, 2008

[30] Auer S, Kuzel HJ, Pollmann H, Sorrentino F. Investigation on MSW fly ash treatment by reactive calcium aluminates and phase formed. Cement and Concrete Research. 1995;**25**:1347-1359

[31] Bournonville B, Nzihou A, Sharrock P, Depelsenaire G. Stabilization of heavy metal containing dusts by reaction with phosphoric acid: Study of the reactivity of fly ash. Journal of Hazardous Materials. 2004;**116**:65-74

[32] Phair JW, Van Deventer JSJ, Smith JD. Effect of Al source and alkali activation on Pb and Cu immobilization in fly ash based geopolymers. Applied Geochemistry. 2004;**19**:423-434

[33] Qian G, Zhang H, Zhang X, Chui P. Modification of MSW fly ash by anionic chelating surfactant. Journal of Hazardous Materials. 2005;**121**:251-258

[34] Yang L. Evaluation of in situ chemical fixation for remediation of arsenic contaminated soil. PhD Dissertation. University of Alabama; 2005

[35] Bhattacharyya S, Donahoe RJ, Patel D. Experimental study of chemical treatment of coal fly ash to reduce the mobility of priority trace elements. Fuel. 2009;**88**:1173-1184

[36] Kim AG, Cardone C. Preliminary Statistical Analysis of Fly Ash Disposal in Mined Areas. Proc: 12th International Symposium on Coal Combustion By-Product Management and Use. American Coal Ash Association; 1997. Vol. 1. pp. 11-1 to 11-13

[37] Kim AG, Kazonich G. Release of Trace Elements from CCB: Maximum Extractable Fraction. Proceedings 14th International Symposium on Management and Use of Coal Combustion Products (CCPs); 2001. Vol. 1. pp. 20-1 to 20-15

[38] Ramezanianpour AA. Fly Ash. Cement Replacement Materials: Properties, Durability, Sustainability. Berlin, Heidelberg: Springer Berlin Heidelberg; 2014. pp. 47-156

[39] American Society for Testing and Materials. Specification for Fly Ash and Raw of Calcined Natural Pozzolan for Use as a Mineral Admixture in Portland Cement Concrete. Philadelphia, PA, ASTM C618-78: ASTM; 1978

[40] Mehta PK. Testing and correlation of fly ash properties with respect to pozzolanic behavior. Palo Alto, CA: Electric Power Research Institute; 1994, Report CS3314

[41] Mehta PK. Pozzolanic and cementitious by-products as mineral admixtures for concrete—A critical review. In: Malhotra VM, editor. Proceedings of 1st International Conference on the Use of Fly Ash, Silica Fume, Slag, and Other Mineral By-products in Concrete, Montebello, PQ, July 31–Aug 5, 1983. Detroit, MI: American Concrete Institute; 1983, Special Publication SP-79

[42] ACI Committee 226. Fly ash in concrete. ACI Materials Journal. 1987;**84**:381-409

[43] Diamond S, Symposium N. Annual Meeting of the Materials Research Society, Boston, MA; 1981. pp. 12-23

[44] Watt JD, Thorne DJ. Composition and pozzolanic properties of pulverized fuel ashes: Composition of fly ashes from some British power stations and properties of their component particles. Journal of Applied Chemistry. 1965;**15**:585-594

[45] Liem H, Sandstrom M, Wallin T, Carne A, Rydevik U, Thurenius B, Moberg P. Studies on the leachate and weathering processes of coal ashes. Water Science and Technology. 1983;**15**:163

[46] Hartlen J, Lundgren T. Disposal of waste material - Evaluation in the KHM project of short and long-term effects. Water Science and Technology. 1983;**15**:49

[47] E1-Mogazi E, Lisk D, Weinstein L. A review of physical, chemical, and biological properties of fly ash and effects on agricultural ecosystems. The Science of the Total Environment. 1988;**74**:1

[48] Cheremisinoff P. Coal flyash: Power plant waste or by-product? Power Engineering. 1988;**92**(7):40

[49] Tazaki K, Fyfe W, Sahu K, Powell M. Observation on the nature of fly ash particles. Fuel. 1989;**68**:727

[50] Electric Power Research Institute (EPRI). Fly Ash Design Manual for Road and Site Applications, Vols. 1 and 2, EPRI TR-100472; 1992, Project 2422-2

[51] Hansen LD, Fisher GL. Elemental distribution in coal fly ash particles. Environmental Science & Technology. 1980;**14**:1111-1117

[52] Jegadeesan G, Al-Abed SR, Pinto P. Influence of trace metal distribution on its leachability from coal fly ash. Fuel. 2008;**87**:1887-1893

[53] Choi SK, Lee S, Song YK, Moon HS. Leaching characteristics of selected Korean fly ashes and its implications for the groundwater composition near the ash mound. Fuel. 2002;**81**:1080-1090

[54] Galbreath KC, Zygarlicke CJ. Formation and chemical speciation of arsenic-, chromium-, and nickel-bearing coal combustion PM 2.5. Fuel Processing Technology. 2004;**85**:701-726

[55] Świetlik R, Trojanowska M, Jóźwiak MA. Evaluation of the distribution of heavy metals and their chemical forms in ESP-fractions of fly ash. Fuel Processing Technology. 2012;**95**:109-118

[56] Eary LE, Rai D, Mattigod SV, Ainsworth CC. Geochemical factors controlling the mobilization of inorganic constituents from fossil fuel combustion residues: II. Review of the minor elements. Journal of Environmental Quality. 1990;**19**:202-214

[57] Vorres KS. Chemistry of mineral matter and ash in coal: An overview. In: Mineral Matter and Ash in Coal, ACS Symposium Series. Washington, DC: American Chemical Society; 1986. pp. 1-8

[58] Ugurlu A. Leaching characteristics of fly ash. Environmental Geology. 2004;**46**:890-895

[59] Fernandez-Turiel JL, de Carvalho W, Cabanas M, Querol X, Lopez-Soler A. Mobility of heavy metals from coal fly ash. Environmental Geology. 1994;**23**:264-270

[60] Merrick D. Trace Elements from Coal Combustion: Emission. London, UK: IEA Coal Research; 1987

[61] Kalogirou E, Themelis N, Samaras P, Karagiannidis A, Kontogianni S. Fly Ash Characteristics From Waste-to-Energy Facilities and Processes For Ash Stabilization. In: Proceedings of ISWA World Congress 2010, Hamburg, Germany, 15-18 November; 2010

[62] Zielinski RA, Foster AL, Meeker GP, Brownfield IK. Mode of occurrence of arsenic in feed coal and its derivative fly ash, Black Warrior Basin, Alabama. Fuel. 2007;**86**:560-572

[63] Yudovich YE, Ketris MP. Arsenic in coal: A review. International Journal of Coal Geology. 2005;**61**:141-196

[64] Cornelis G, Johnson CA, Gerven TV, Vandecasteele C. Leaching mechanisms of oxyanionic metalloid and metal species in alkaline solid wastes: A review. Applied Geochemistry. 2008;**23**:955-976

[65] Hassett DJ, Pflughoeft-Hassett DF, McCarthy GJ. Ettringite formation in coal ash as a mechanism for stabilization of hazardous trace elements. Proc: 9th Int. Ash Use Symp. 2, 31-1 to 31-17; 1991

[66] Lecuyer I, Bicocchi S, Ausset P, Lefevre R. Physico-chemical characterization and leaching of desulfurization coal fly ash. Waste Management and Research. 1996;**14**:15-28

[67] Rajan SSS. Adsorption and desorption of sulfate and charge relationships in allophonic clays. Soil Science Society of America Journal. 1979;**43**:65-69

[68] Jankowski J, Ward CR, French D. Preliminary Assessment of Trace Element Mobilization from Australian Fly Ashes. Co-operative Research Centre for Coal in Sustainable Development, Research Report 45; 2004. 44 pp. http://pandora.nla.gov.au/pan/64389/20080828-1328/www.ccsd.biz/publications/425.html

[69] Izquierdo M, Querol X. Leaching behaviour of elements from coal combustion fly ash: An overview. International Journal of Coal Geology. 2012;**94**:54-66

[70] Dreesen DR, Gladney ES, Owens JW. Comparison of levels of trace elements extracted from fly ash and levels found in effluent waters from a coal-fired power plant. Environmental Science and Technology. 1977;**11**:1017-1019

[71] Hassett DJ, Pflughoeft-Hassett DF, Heebink LV. Leaching of CCBs: Observations from over 25 years of research. Fuel. 2005;**84**:1378-1383

[72] Nidheesh PV, Singh TSA. Arsenic removal by electrocoagulation process: Recent trends and removal mechanism. Chemosphere. 2017;**181**:418-432

[73] IPCS (International Program on Chemical Safety) and WHO (World Health Organization). Environmental Health Criteria for Arsenic and Arsenic Compounds. Geneva: GreenFacts; 2001

[74] Henke KR, Hutchison A. Arsenic chemistry. In: Arsenic: Environmental Chemistry, Health Threats and Waste Treatment. Hoboken, New Jersey: John Wiley & Sons, Ltd; 2009. pp. 9-68

[75] Henke KR. Arsenic in natural environments. In: Arsenic: Environmental Chemistry, Health Threats and Waste Treatment. Hoboken, New Jersey: John Wiley & Sons, Ltd; 2009. pp. 69-235

[76] IARC Working Group on the Evaluation of Carcinogenic Risks to Humans. Arsenic, metals, fibres, and dusts. IARC Monographs on the Evaluation of Carcinogenic Risks to Humans. 2012;**100**:11-465

[77] Jiang JQ. Removing arsenic from groundwater for the developing world-a review. Water Science and Technology. 2001;**44**:89-98

[78] Singh R, Singh S, Parihar P, Singh VP, Prasad SM. Arsenic contamination, consequences and remediation techniques: A review. Ecotoxicology and Environmental Safety. 2015;**112**:247-270

[79] Chung JY, Yu S, Do Hong YS. Environmental source of arsenic exposure. Journal of Preventive Medicine and Public Health. 2014;**47**:253-257

[80] Al-Abed SR, Jegadeesan G, Purandare J, Allen D. Arsenic release from iron rich mineral processing waste: Influence of pH and redox potential. Chemosphere. 2007;**66**:775-782

[81] Duarte AALS, Cardoso SJA, Alçada AJ. Emerging and innovative techniques for arsenic removal applied to a small water supply system. Sustainability. 2009;**1**:1288-1304

[82] Song S, Gallegos-Garcia M. Arsenic removal from water by the coagulation process. In: Fanun M, editor. The Role of Colloidal Systems in Environmental Protection. Elsevier; 2014. pp. 261-277

[83] Jay Murray F. A human health risk assessment of boron (boric acid and borax) in drinking water. Regulatory Toxicology and Pharmacology. 1995;**22**:221-230

[84] Sahin S. A mathematical relationship for the explanation of ion exchange for boron removal. Desalination. 2002;**143**:35-43

[85] Tu KL, Ngheim LD, Chivas AR. Boron removal by reverse osmosis membranes in seawater desalination. Separation and Purification Technology. 2010;**75**:87-101

[86] Power PP, Woods WG. The chemistry of boron and its speciation in plants. Plant and Soil. 1997;**193**:1-13

[87] Isa MH, Ezechi EH, Ahmed Z, Magram SF, Kutty SRM. Boron removal by electrocoagulation and recovery. Water Research. 2014;**51**:113-123

[88] Wyness AJ, Parkaman RH, Neal C. A summary of boron surface water quality data throughout the European Union. Science of the Total Environment. 2003;**314-316**:255-269

[89] Wolska J, Bryjak M. Methods for boron removal from aqueous solutions — A review. Desalination. 2013;**310**:18-24

[90] RACI (Royal Australian Chemical Institute). 2011. Selenium. Available at: www.raci.org.au

[91] Sandy T, DiSante C. Review of Available Technologies for the Removal of Selenium from Water. Englewood, Colorado: CH2M HILL Inc; 2010. pp. 2-223

[92] Moore L, Mahmoudkhani A. Methods for removing selenium from aqueous systems. In: Proceedings Tailings and Mine Waste 2011, Vancouver, BC, November 6 to 9; 2011

[93] Sabbas T, Polettini A, Pomi R, Astrup T, Hjelmar O, Mostbauer P, Cappai G, Magel G, Salhofer S, Speiser C. Management of municipal solid waste incineration residues. Waste Management. 2003;**23**:61-88

[94] Fruergaard T, Astrup T. Life cycle assessment of management of APC residues from waste incineration. In: Proceedings Sardinia 2007, Eleventh International Waste Management and Landfill Symposium, Sardinia, Italy, 1-5 October; 2007

[95] Lundtorp K. The Ferrox-process in an industrial scale - Developing a stabilization process for air pollution control residues from municipal solid waste incineration. PhD Thesis. Technical University of Denmark, Environment & Resources DTU; 2001

[96] Hong KJ, Tokunaga S, Ishigami Y, Kajiuchi T. Extraction of heavy metals from MSW incinerator y ash using saponins. Chemosphere. 2000;**41**:345-352

[97] Bosshard PP, Bachofen R, Brandl H. Metal leaching of y ash from municipal waste incineration by *Aspergillus niger*. Environmental Science and Technology. 1996;**30**:3066-3070

[98] Pedersen AJ, Ottosen LM, Villumsen A. Electrodialytic removal of heavy metals from different fly ashes - Influence of heavy metal speciation in the ashes. Journal of Hazardous Materials. 2003;**100**:65-78

[99] Crillesen K. Information about the "Askepot" Project, Personal Communication. Denmark: I/S Vestforbraending; 2005

[100] Quina MJ, Bordado JC, Quinta-Ferreira RM. Treatment and use of air pollution control residues from MSW incineration: An overview. Waste Management. 2008;**28**:2097-2121

[101] Gong B, Deng Y, Yang Y, He Y, Sun X, Ge L-Y, et al. Stabilization of lead in incineration fly ash by moderate thermal treatment with sodium hydroxide addition. PLoS One. 2017;**12**(6):e0178816

[102] Chi M. Effects of modulus ratio and dosage of alkali-activated solution on the properties and microstructural characteristics of alkali-activated fly ash mortars. Construction and Building Materials. 2015;**99**:128-136

[103] Lundtorp K, Jensen DL, Christensen TH. Stabilization of APC residues from waste incineration with ferrous sulfate on a semi-industrial scale. Journal of the Air & Waste Management Association. 2002;**52**:722-731

[104] Ecke H, Menad N, Lagerkvist A. Carbonation of municipal solid waste incineration fly ash and the impact on metal mobility. Journal of Environmental Engineering. 2003:435-440

[105] Jiang J, Chen M, Zhang Y, Xu X. Pb stabilization in fresh fly ash from municipal solid waste incinerator using accelerated carbonation technology. Journal of Hazardous Materials. 2009;**161**(2-3):1046-1051

[106] Hjelmar O, Birch H, Hansen JB. Development of a process for treatment of air pollution control residues from MSW incinerators prior to landfilling. In: Christensen TH, Cossu R, Stegmann R, editors. Seventh International Waste Management and Landfill Symposium, 543-548. Sardinia, CISA, Italy. 1999

[107] Kim S, Matsuto T, Tanaka N. Evaluation of pre-treatment methods for landfill disposal of residues from municipal solid waste incineration. Waste Management & Research. 2003;**21**:416-423

[108] Youcai Z, Lijie S, Guojian L. Chemical stabilization of MSW incinerator fly ashes. Journal of Hazardous Materials. 2002;**B95**:47-63

[109] Vandecasteele C, Dutr'e V, Geysen D, Wauters G. Solidification/stabilisation of arsenic bearing fly ash from the metallurgical industry. Immobilisation Mechanism of Arsenic, Waste Management. 2002;**22**:143-146

[110] Dutre V, Vandecasteele C. Immobilization of arsenic in waste solidified using cement and lime. Environmental Science & Technology. 1998;**32**:2782-2787

[111] Zacco A, Gianoncelli R, Ardesi S, Sacrato LG, Bontempi E. Use of colloidal silica to obtain a new inert from municipal solid waste incinerator (MSWI) fly ash: First results about reuse. Clean Technologies and Environmental Policy. 2012;**14**(2):291-297

[112] Rodella N, Bosio A, Zacco A, Borgese L, Pasquali M, Dalipi R, Depero LE, Patel V, Bingham PA, Bontempi E. Arsenic stabilization in coal fly ash through the employment of waste materials. Journal of Environmental Chemical Engineering. 2014;**2**:1352-1357

[113] Conner JR, Hoeffner SL. A critical review of stabilization/solidification technology. Critical Reviews in Environmental Science and Technology. 1998;**28**:397-462

[114] Todorovic J, Ecke H, Lagerkvist A. Solidification with water as a treatment method for air pollution control residues. Waste Management. 2003;**23**:621-629

[115] IAWG (International Ash Working Group), Chandler AJ, Eighmy TT, Hartlen J, Hjelmar O, Kosson DS, Sawell SE, van der Sloot HA, Vehlow J. Municipal Solid Waste Incinerator Residues. Studies in Environmental Science. Amsterdam: Elsevier Science; 1997. p. 67

[116] Margallo M, Taddei MBM, HernaÂndez-PelloÂn A, Aldaco R, Irabien AÂ. Environmental sustainability assessment of the management of municipal solid waste incineration residues: A review of the current situation. Clean Technologies and Environmental Policy. 2015;**17**(5):1333-1353

[117] Lindberg D, Molin C, Hupa M. Thermal treatment of solid residues from WtE units: A review. Waste Management. 2015;**37**:82-94

[118] Sakai S, Hiraoka M. Municipal solid waste incinerator residue recycling by thermal processes. Waste Management. 2000;**20**:249-258

[119] Wey MY, Liu KY, Tsai TH, Chou JT. Thermal treatment of the fly ash from municipal solid waste incinerator with rotary kiln. Journal of Hazardous Materials. 2006;**137**(2):981-989

[120] Ecke H, Sakanakura H, Matsuto T, Tanaka N, Lagerkvist A. State-of-the-art treatment processes for municipal solid waste incineration residues in Japan. Waste Management and Research. 2000;**18**:41-51

[121] Hartuti S, Kambara S, Takeyama A, Kumabe K, Moritomi H. Direct quantitative analysis of arsenic in coal fly ash. Journal of Analytical Methods in Chemistry. 2012;**2012**:438,701

[122] Cervera ML, Arnal MC, de la Guardia M. Removal of heavy metals by using adsorption on alumina or chitosan. Analytical and Bioanalytical Chemistry. 2003;**375**:820-825

[123] Su T, Guan XH, Gu GW, Wang JM. Adsorption characteristics of As(V), Se(IV) and V(V) onto activated alumina: Effects of pH, surface loading, and ionic strength. Journal of Colloid and Interface Science. 2008;**326**:347-353

[124] Cornell RM, Schwertmann U. The Iron Oxides: Structure, Properties, Reactions, Occurrences and Uses. 2nd ed. Weinheim, Germany: Wiley-VCH; 2003

Evaluation of Ultrasound-assisted Modified Fly Ash for Treatment of Acid Mine Drainage

Deniz Sanliyuksel Yucel and Burcu Ileri

Additional information is available at the end of the chapter

http://dx.doi.org/10.5772/intechopen.69425

Abstract

The aim of this chapter is to assess the suitability of utilizing fly ash (FA) and modified fly ash (MFA) in the treatment of acid mine drainage (AMD). The effects of various experimental conditions such as mixing speed, fly ash dosage, contact time, and pH on the efficiency of metal removal from AMD was studied in batch experiments. For modification of FA, ultrasound (US) process was conducted at low frequency (20 kHz) to activate surface area and decrease particle size of FA at 90 W for 30 minutes. Chemical, physical and mineralogical compositions of FA, MFA and solid residues (SR) were determined using arrays of X-ray fluorescence (XRF), Brunauer-Emmett-Teller (BET), scanning electron microscopy coupled with energy dispersive X-ray spectrometry (SEM-EDX), X-ray diffraction (XRD), Fourier transform infrared spectroscopy (FTIR) techniques and particle size analysis. The removal efficiency of Al, Fe, Mn, Ni, Zn and SO_4^{2-} from AMD was compared using FA and MFA. According to the obtained results, ultrasound-assisted MFA provided successful results for metal removal in a shorter reaction time and with smaller dosage than FA. As the pH of the solution (ash:AMD) increased, the metal and SO_4^{2-} concentrations in SR increased mainly due to precipitation; this data was verified with XRD and XRF analysis. Preliminary treatment of AMD from Etili coal mine (NW Turkey) indicates that MFA could be an effective and low-cost adsorbent for the treatment of AMD.

Keywords: acid mine drainage, modified fly ash, ultrasound, metal, sulphate, solid residue

1. Introduction

The Can coal basin (Canakkale province, NW Turkey) is rich in lignite (100 Mt); though these reserves, known as Can coals, cannot be described as quality coals due to low calorific value and high sulphur content (max: 14.36 wt%) [1]. In the Can coal basin, the Etili mine has operated as open pit since the 1980s [2]. When mining activities ceased without any accompanying remediation methods, these depressions filled with surface and ground waters. During the time, artificial lakes formed and showed typical characteristics of acid mine lakes (AML) with low pH (min: 2.59) and high electrical conductivity values (max: 6430 μS/cm), in addition to high SO_4^{2-} (max: 5370 mg/L) and metal (Al, Fe, Mn, etc.) concentrations [3]. Acid mine lakes are located upstream of the Kocabas stream, which passes through the Can coal basin before flowing into the Sea of Marmara. Along this stream, drinking water for several residential areas is supplied from boreholes [4]. The pH value of creeks around the mine lakes varies between 2.85 and 5.75 [5], with these creeks passing close to settlement areas. Soil and groundwater quality has been affected along these creeks. These conditions prohibit discharge of untreated acid mine drainage (AMD) into public streams, as they have a detrimental effect on environment. So, for removal of metals from AMD, suitable physical or chemical treatment processes such as chemical precipitation, adsorption, oxidation-reduction, ion exchange, etc. must be performed [6–12]. Adsorption refers to a mass transfer process by which a substance is transferred from the liquid phase to the surface of a solid and becomes bound by physical and/or chemical interactions [13]. Activated alumina, activated carbon, silica gel, fly ash, saw dust and zeolite are different adsorbent materials reported in the literature [10–16] for removal of pollutants from water with the adsorption process.

Fly ash (FA), which is a low cost and abundant industrial by-product, has been known for its effectiveness in the remediation of different kinds of industrial wastes in Europe and USA since 1960s [14]. In Turkey, 13 million tons of FA is produced annually and this amount is expected to reach 50 million tons per year by 2020 [17, 18]. But only a small amount of FA in Turkey (approximately 1%) is reused in the construction industry (particularly cement industry) [19, 20] and the rest is disposed of in ash dumps or landfill which is an inconvenient solution both from environmental and economical points of view [21]. Recent studies have demonstrated that treatment of AMD with FA is an effective adsorption process for metal uptake from water [16, 22–28]. Ultrasound (US) has been used recently as one of the most successful means of activating adsorbate surfaces [29, 30]. The US technology works according to the principle of cavitation and the adsorbent surface modification is achieved by means of pyrolysis and •OH radicals which burst in the water under high temperature and pressure and this results in continuity of use of the adsorbent for purification purposes [30–33]. High power US produces a permanent chemical or physical change in a material and •H, •OH and $\bullet OH_2$ radicals when water is liquid irradiated with US [34–36]. High intensity shear forces and radicals formed during the US mechanical effect create pits and cracks on adsorbate surfaces resulting in larger specific surface area, higher pore volume and lower particle diameter [37]. The use of US cavitation with adsorption for water treatment has been reported in the literature [38–40]. This chapter is a site-specific research carried out to assess the suitability of utilizing FA and modified fly ash (MFA) in the treatment of AMD from Etili coal mine. Within this chapter, an AML that is one of the largest mine lakes in Etili coal mine 1 km Southwest of Keciagili village with 29.41 decar area in July 2016 not drained since the 1990s was chosen (**Figure 1a** and **b**). Fly ash was collected directly from the fluidized bed combustion thermal power plant precipitator, which is a coal-fired 135 MW generating facility, located in Northwestern Turkey. In this chapter, FA, a coal combustion by-product, was US-assisted modified with the aim of activating the surface area and decreasing particle size. Removal of metal and SO_4^{2-} using FA and MFA was determined under various conditions such as mixing speed, ash

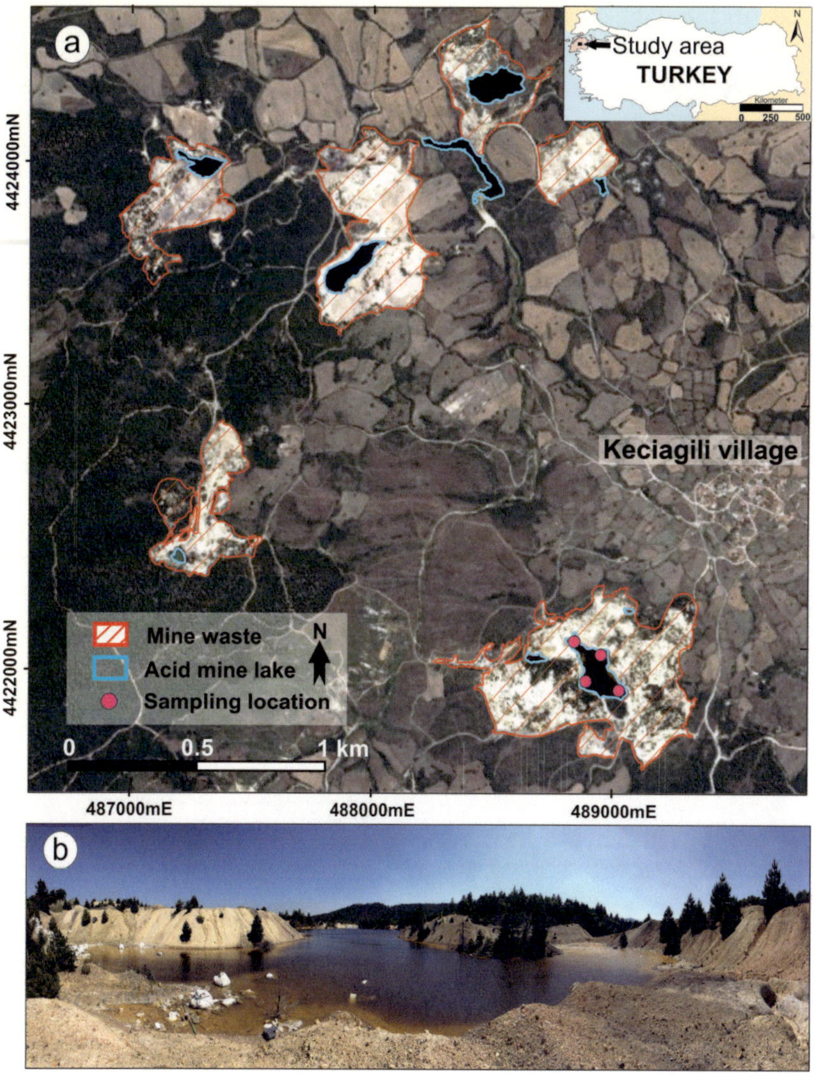

Figure 1. (a) Location map of Etili coal mine (b) snapshot from sampling AML.

dosage, contact time and pH in batch experiments. The possibility of utilizing FA and MFA to remove selected metals and SO_4^{2-} from AMD is evaluated.

2. Materials and methods

2.1. Acid mine lake characterization

Four different sampling points were selected and water samples were taken from 0.5 to 1 m depths in the lake using a bailer sampler (Eijkelkamp). Physicochemical parameters of AML were measured in the field, including pH, oxidation-reduction potential (ORP, mV), temperature (T, °C), electrical conductivity (EC, μS/cm), dissolved oxygen (DO, mg/L), total dissolved solids (TDS, mg/L) and salinity (Sal, %), using a multi-parameter measurement device (WTW 3410). All water samples were stored in 500-mL polyethylene bottles and refrigerated at 4°C until further analysis. Water samples were analysed for major and trace elements using an inductively coupled plasma-mass spectrometer (ICP-MS) at AcmeLabs (Canada). Acidity was measured as methyl orange acidity with digital titrator (APHA 2310 titration method), and Fe(II) was determined by ferrozine spectrophotometric method using a Hach Lange DR 2800 spectrophotometer. Fe(III) was calculated as the difference between Fe(t) and Fe(II). Sulphate was measured by Hach Lange DR 2800 using SulfaVer 4 method.

2.2. Fly ash modification

The FA modification was achieved using US technology. The US system was equipped with a titanium probe having diameter of 19 mm, connected to a horn-transducer of 20 kHz supported with a booster and eventually plugged to a power supply of 500 W with a cable (VibraCell 505, Sonics&Materials) (**Figure 2**). The batch reactor had a cooling jacket through which tap water was circulated to absorb the heat generated by US treatment. Each 200 mL pure water sample initially containing 10 g of FA was placed in the cooling-jacket reactor, and then modified with US (20 kHz) at 90 W for 10, 30 and 60 minutes. Modified FA was dried in the oven at 80°C for 12 hours. The variations in particle size of MFA at 10, 30 and 60 minutes were determined to identify optimum modification time according to particle size analysis

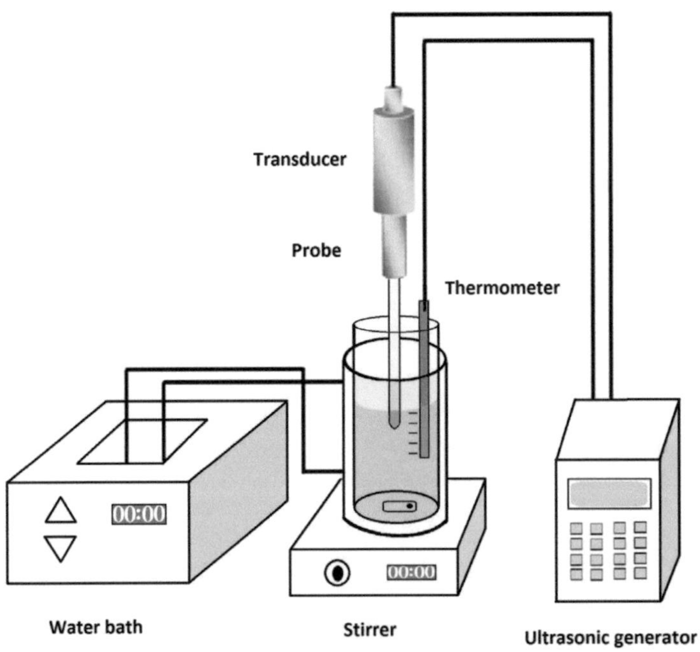

Figure 2. Low frequency ultrasonic system (20 kHz).

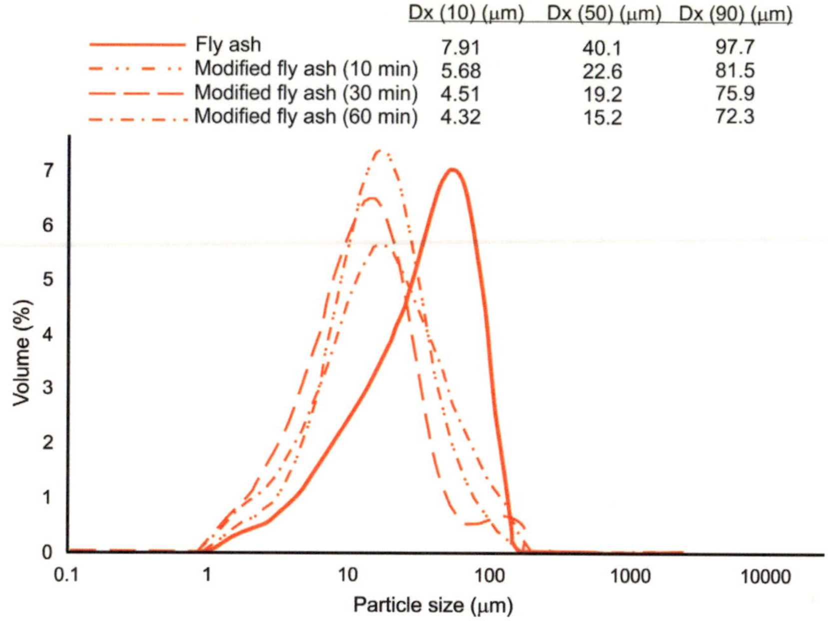

Figure 3. Particle size distribution of FA and MFA.

(**Figure 3**). The particle size diameter of FA was decreased and surface area was increased by US mechanical effects. The obtained particle size distribution resulted for 30 minutes had similar Dx (10, 50 and 90) values compared to 60 minutes. Therefore, 30 minutes was chosen as optimum time for modification of FA due to reduce US energy, cost and time.

2.3. Batch experiments

Batch experiments were conducted to determine various parameters such as (1) mixing speed (rpm), (2) ash dosage (g), (3) contact time (hour) and (4) pH that affect the removal of metal and sulphate (SO_4^{2-}) from AMD using FA and MFA at ambient temperature. (1) Twenty grams of FA and MFA was mixed with 100 mL AMD samples and was agitated under different mixing speeds (60, 120 and 180 rpm) using an orbital shaker (GFL®3005) to determine optimum mixing speed for 0–6 hours contact time. It was found that removal of metal increased with increasing mixing speed. Thus, 180 rpm was selected for optimum mixing speed. (2) FA and MFA were weighed at 5, 10 and 20 g dosage and subsequently added to AMD of 100 mL volume in a conical flask. The mixed sample was agitated using an orbital shaker at 180 rpm for 0–6 hours to determine optimum adsorbent dosage for Fe removal. (3) The effect of contact time on the adsorption of metal (Al, Fe, Mn, Ni and Zn) and SO_4^{2-} from AMD using 10 and 20 g FA and MFA dosage was evaluated with various contact times (0.2, 0.5, 1, 2, 3, 4, 5 and 6 hours) at 180 rpm. (4) The pH of the AMD was adjusted to 3, 5, 7 and 9 using 1 N NaOH before the tests. All experiments were conducted according to optimum mixing speed (180 rpm), 20 g adsorbent dosage, 6 hour contact time and various pH values to determine Al, Fe, Mn, Ni, Zn and SO_4^{2-} removal from the solution. At the end of mixing, FA and MFA were separated from suspension by filtration through a 0.45 μm membrane filter. pH and EC values of the suspension were measured with a multi-parameter device. The residual concentrations of the metal ions in the supernatant were determined using ICP-MS technique. The amount of metal ions adsorbed by FA and MFA was taken as the difference between the initial and residual concentration of the metal ions. Each test was repeated at least three times. In each case, the percentage error for the experiment was lower than ±5%. The solid residues (SR) including FA:AMD and MFA:AMD determined for optimum Fe removal results could settle for 5 minutes and separated from water. Then, SR was dried in the oven at 80°C for 12 hours to determine SR characterization.

2.4. Fly ash and modified fly ash characterization

The physical, mineralogical and chemical composition of FA, MFA and SR were identified with particle size, Brunauer-Emmett-Teller (BET), scanning electron microscopy (SEM) coupled with energy-dispersive X-ray spectrometry (EDX), X-ray diffraction (XRD), Fourier transform infrared spectroscopy (FTIR) and X-ray fluorescence (XRF) analyses. The particle size distribution of FA, MFA and SR were measured using a laser-based particle size analyser (PSA), namely a Malvern Mastersizer 3000 PSA, in the Polymer Laboratory at Chemistry Department in Canakkale Onsekiz Mart University. Morphology, particle shape and mineralogical components of FA, MFA and SR were examined through micro-graphs taken by SEM (JEOL SEM-7100-EDX) coupled with EDX performed at the Centre for Material Science in Canakkale Onsekiz Mart University. Specific surface area of FA and MFA was identified by BET (Quantachrome, Quadrasorb SI) analyses at the Centre for Material Science in Canakkale Onsekiz Mart University. Mineralogical compositions of FA, MFA and SR were identified by XRD using Philips PW 1830 instrument at the General Directorate of Mineral Research and Exploration Analysis Laboratories in Ankara, Turkey. Both major and minor element concentrations of FA, MFA and SR were obtained using an XRF spectrometer (PANalytical, Axios) at the General Directorate of Mineral Research and Exploration Analysis Laboratories. The chemical compounds of FA, MFA and SR were identified using a single-beam FTIR-PerkinElmer spectrum 100 spectrometer, equipped with a diamond crystal in the Chemistry Laboratory in Canakkale Onsekiz Mart University. IR spectra were recorded as changes in absorption as a function of wave number ranging from 600 to 4000 cm^{-1}.

3. Results and discussions

3.1. Hydrochemical characteristics of acid mine lake

The pH value of the mine lake was strongly acidic (2.66). Low pH value of the lake increases the dissolution of numerous elements. Electrical conductivity value of the mine lake was measured as 5.2 mS/cm (**Table 1**). Electrical conductivity is related to the concentration of TDS and major ions. Total dissolved solids may be obtained by multiplying the conductance by a factor, which is commonly between 0.55 and 0.75, and generally multiplication factor is approximately 0.67 [41]. Total dissolved solids were measured and calculated as 2720 and

Parameter	Unit	Concentration
T	°C	26.3
pH	-	2.66
ORP	mV	355
EC	mS/cm	5.2
Sal	%	2.7
TDS	mg/L	2720
DO		5.7
Acidity		2000
Ca^{2+}		496
Mg^{2+}		267
Na^{+}		156
K^{+}		3
SO_4^{2-}		3600
Cl^{-}		65
Al		249.9
Fe(t)		199.7
Fe^{2+}		37
Fe^{3+}		162.7
Mn		114.74
Zn		6.21
Co		2.29
Ni		1.11
Cu		0.23

Table 1. Hydrochemical characteristics of AML.

3484 mg/L, respectively. Temperature of the lake (26.3°C) was related to climatic fluctuations (e.g. season and different times of day and depth of water body). Oxidation-reduction potential and DO values were measured as 355 mV and 5.7 mg/L, respectively. Oxidation-reduction potential influences the mobility of most metals, which can exist in several oxidation states. Typically, high ORP values correspond to the presence of oxidizing agents. Thus, ORP values are typically found to be high when DO levels are high [42]. Acidity values based on titrations performed in the laboratory were quite high, 2000 mg$CaCO_3$/L. The titrated values are slightly lower than acidity values (2128 mg$CaCO_3$/L) calculated using the formula of [43]. Major cation order is Ca^{2+} (496 mg/L) > Mg^{2+} (267 (mg/L) > Na^{+} (156 mg/L) > K^{+} (3 mg/L), and major anion order is SO_4^{2-} (3600 mg/L) > Cl^{-} (65 mg/L). Therefore, lake water is best classified as Ca-SO_4 water. The concentrations of Fe and Al in the mine lake are in most cases greater than those of the common ions Na^{+} and K^{+}. The dominant dissolved metals in the lake water, in order of average abundance, are Al (249.9 mg/L)> Fe (199.7 mg/L) > Mn (111.74 mg/L) > Zn (6.21 mg/L) > Co (2.29 mg/L) > Ni (1.11 mg/L) > Cu (0.23 mg/L). Most the dissolved Fe in the lake water is Fe^{3+} (162.7 mg/L). According to the Turkish inland water quality regulations [44], the mine lake had pH, SO_4^{2-}, Al, Fe, Mn, Zn, Co, Ni and Cu values in class IV, and were classified as highly polluted waters.

3.2. Mineralogical, physical and chemical characterization of original and modified fly ash

The mineralogical, physical and chemical properties of fly ash depend on the nature of the parent coal, conditions of combustion, type of emission control devices, storage and handling methods [45, 46]. Fly ash (pH = 10.95) and MFA (pH = 11.58) have alkaline character when added to deionized water 1:2 (solid:solution ratio) at the end of 24 hours (**Table 2**). Electrical conductivity value of MFA (2.24 mS/cm) is slightly lower than EC value of FA (2.35 mS/cm). The chemical composition of the FA was determined by XRF analyses to contain 50.75% SiO_2 + 19.98% Al_2O_3 + 5.37% MgO + 5.35% Fe_2O_3 + 2.15% Na_2O and 1.81% CaO and 10.9% LOI. According to the American Society for Testing Materials [47], the ash is classified as 'F-class fly ashes', the sum of SiO_2 + Al_2O_3 + Fe_2O_3 > 70% and CaO content is less than 5%. The combustion of bituminous or anthracite coal produces F-class FA and F-class FA is pozzolanic [48]. The chemical composition of MFA contains 50.48% SiO_2 + 19.59% Al_2O_3 + 5.58% MgO + 5.39% Fe_2O_3 + 2.05% Na_2O and 1.74% CaO and 11.65% LOI. Specific surface area (BET) of FA is increased with US-assisted modification from 3.21 to

Parameter	Unit	FA	MFA
pH	-	10.95	11.58
EC	mS/cm	2.35	2.24
Specific surface area	m^2/g	3.21	4.54
SiO_2	wt%	50.75	50.48
Al_2O_3		19.98	19.59
MgO		5.37	5.58
Fe_2O_3		5.35	5.39
Na_2O		2.15	2.05
CaO		1.81	1.74
K_2O		1.75	1.72
MnO		0.052	0.051
ZnO		0.017	0.017
NiO		0.008	0.008
CuO		0.006	0.006
LOI		10.90	11.65

Table 2. Chemical composition of FA and MFA.

4.54 m^2/g. The MFA contains lower amounts of Si and Al and has a higher surface area, LOI and pH than the FA. The results for the mineralogical composition of FA and MFA remained the same as obtained by XRD analysis. Mineralogical composition of FA includes quartz (42%), amorphous phase (23%), mullite (13%), plagioclase (12%), hematite (6%), magnetite (2%), mica (1%) and K-feldspar (1%). Modified FA contains quartz (37%), amorphous phase (31%), mullite (10%), plagioclase (10%), mica (5%), K-feldspar (4%), hematite (2%) and magnetite (1%). Ultrasound-assisted modification decreased particle size diameter of FA and so the percentages of mineral composition changed. Particle size characterization of FA is important for metal adsorption. Adsorption capacity of the smallest particles is better as they are more reactive, and in addition trace elements tend to concentrate in the finest particles. The mean particle size distribution of the FA and MFA is presented in **Figure 3**. The results show that the particle size distribution of the FA and MFA ranges between 1 and 144 μm and 0.9 and 116 μm, respectively. From the cumulative distributions, the following values for Dx (10, 50 and 90) of FA and MFA were observed 7.91, 40.1 and 97.7 μm and 4.51, 19.2 and 75.9 μm, respectively. The size and shape of the FA were scanned with SEM-EDX as shown in **Figure 4a–d**. Generally, FA particles have an irregular and porous microscopic structure. Most of the FA particles exhibit irregular shapes, primarily because most minerals in the coal do not undergo melting but soften only, under the relatively low boiler temperature of 850–900°C [49–51]. Irregularly shaped unburned carbon particles tended to be in the upper end of the size distribution. The occurrence of hematite results from oxidation of pyrite in the combustion units. Fly ash and MFA contain small pores, which range between 0.6 and 11.2 μm. Ultrasound mechanical effect created pits and cracks on FA surface and part of ash particles filled inside the pores (**Figure 5a–d**).

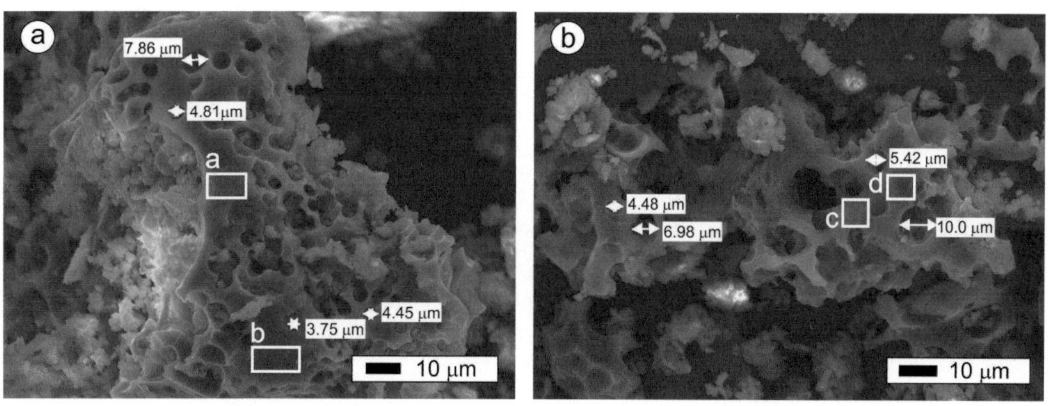

(c)

	a (At, %)	b (At, %)
C	99.5	99.3
Si	0.3	0.4
Al	0.1	0.2
S	0.1	0.1

(d)

	c (At, %)	d (At, %)
C	99.1	99.2
Si	0.5	0.4
Al	0.3	0.4
S	0.1	-

Figure 4. (a, b) SEM images of FA, (c, d) EDX results of FA.

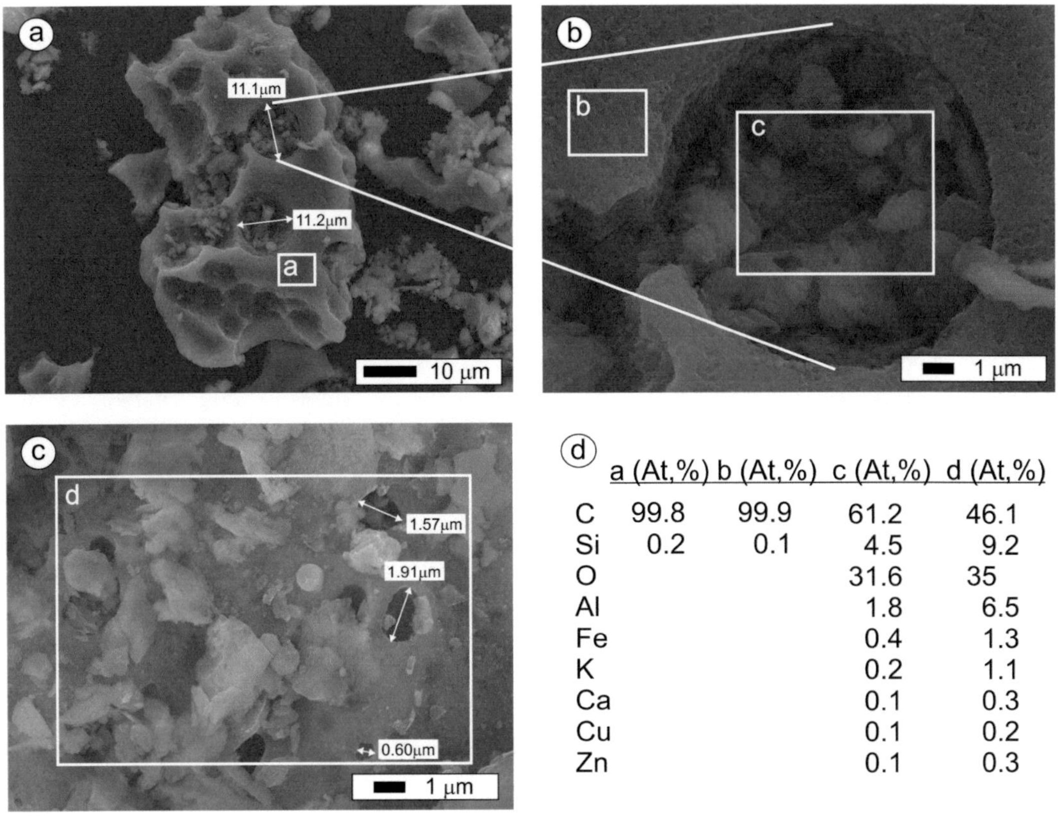

	a (At,%)	b (At,%)	c (At,%)	d (At,%)
C	99.8	99.9	61.2	46.1
Si	0.2	0.1	4.5	9.2
O			31.6	35
Al			1.8	6.5
Fe			0.4	1.3
K			0.2	1.1
Ca			0.1	0.3
Cu			0.1	0.2
Zn			0.1	0.3

Figure 5. (a–c) SEM images of MFA, (d) EDX results of MFA.

3.3. Metal removal from acid mine drainage using fly ash and modified fly ash

3.3.1. *Effect of agitation*

At higher agitation speeds, the mass transfer resistance related to ion diffusion through the liquid film surrounding the FA is reduced as the film thickness reduces due to agitation, resulting in greater metal uptake [52]. To determine the effect of agitation on the removal of Fe (t), AMD (100 mL) was homogeneous mixed with FA and MFA (20 g) at 60, 120 and 180 rpm for 0–6 hours. The results of the Fe concentration vs. contact time are shown in **Figure 6a** and **b**. According to the obtained results, the speed of agitation is an important factor for removal of Fe from solution. The maximum removal of Fe increased with increasing mixing speed and the optimum mixing speed was determined as 180 rpm at least time. The Fe concentration of the solution was measured 4.6, 0.38 and 0.1 mg/L; 4.5, 0.30 and 0.05 mg/L for 60, 120 and 180 rpm at the end of 6 hours for FA:AMD and MFA:AMD, respectively.

3.3.2. *Effect of adsorbent dosage*

Neutralization is generally the first step in AMD treatment [11]. The kinetics of the neutralization reactions was investigated by monitoring the pH and EC values of FA:AMD and

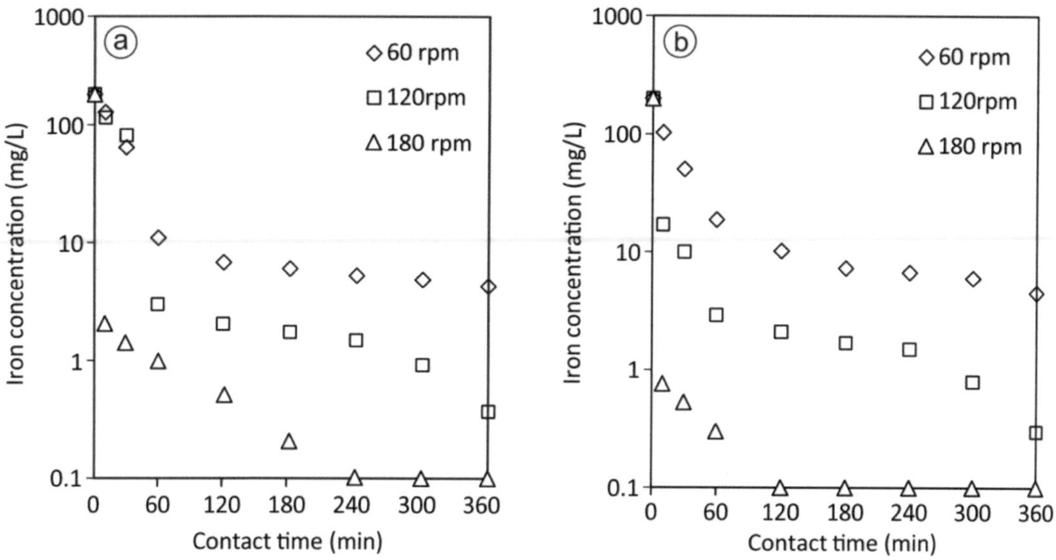

Figure 6. Effect of agitation on Fe concentration of AMD mixing with (a) FA and (b) MFA.

MFA:AMD ratio (1:20, 1:10 and 1:5) over a period of 6 hours. Initial pH (2.66) of the FA:AMD increased within the first 30 minutes for all ash dosages, indicating that soluble material dissolved quickly resulting in the solution. pH values reached 4.28, 4.43 and 5.18 for 5, 10 and 20 g FA dosage at the end of 6 hours, respectively (**Figure 7a**). Electrical conductivity values ranged from 5.2 to 4.8 mS/cm, 5.2 to 4.81 mS/cm and 5.2 to 4.51 mS/cm were measured for the 5, 10 and 20 g FA dosage at 0–6 hours, respectively (**Figure 7b**). pH value of the solution reached 5.51, 7.3 and 7.9 for 5, 10 and 20 g MFA dosage at the end of 6 hours (**Figure 7c**). Low pH value with a minimum of 2.66 was measured in initial, and the highest pH value was measured ash dosage of 20 g at 6 hours, which compared well with the adsorbent dosage and contact time. Alkaline pH value of solution indicates that modification of FA affects the alkalinity. Acid mine drainage reacted with MFA in 6 hours using 1:5 and 1:10 MFA:AMD ratios by weight produces neutral and alkaline process waters. Electrical conductivity values ranged between 5.2 and 4.85 mS/cm, 5.2 and 4.8 mS/cm, 5.2 and 4.2 mS/cm were measured for MFA dosage of 5, 10 and 20 g at 0–6 hours, respectively (**Figure 7d**). A sudden increase in EC was observed during the first 10 minutes for all MFA dosages, indicating that soluble material dissolved into the solution resulting in a rapid and irregular increase in EC value. Adsorbent dosage is effective parameter for metal removal [53]. In order to determine the effect of adsorbent dosage on the removal of Fe, AMD (100 mL) was homogeneously mixed with FA and MFA (5, 10 and 20 g) at 180 rpm for 0–6 hours. According to FA, various adsorbent dosages (5, 10 and 20 g), Fe concentration were determined 1.43 mg/L at pH = 4.28, 1.03 mg/L at pH = 4.43, 0.1 mg/L at pH = 5.18 at the end of 6 hours, respectively (**Figure 8a**). According to Turkish inland water quality regulations [44], Fe concentration must be lower than 0.3 mg/L for unpolluted water classification (Class I). Considering this regulation, optimum FA dosage for Fe removal from AMD was determined as 20 g in this chapter. Fe concentration shows approximately 99.9% removal at pH = 5.18 attained by the FA:AMD ratio of 1:5. According to various adsorbent dosages (5, 10 and 20 g) for MFA,

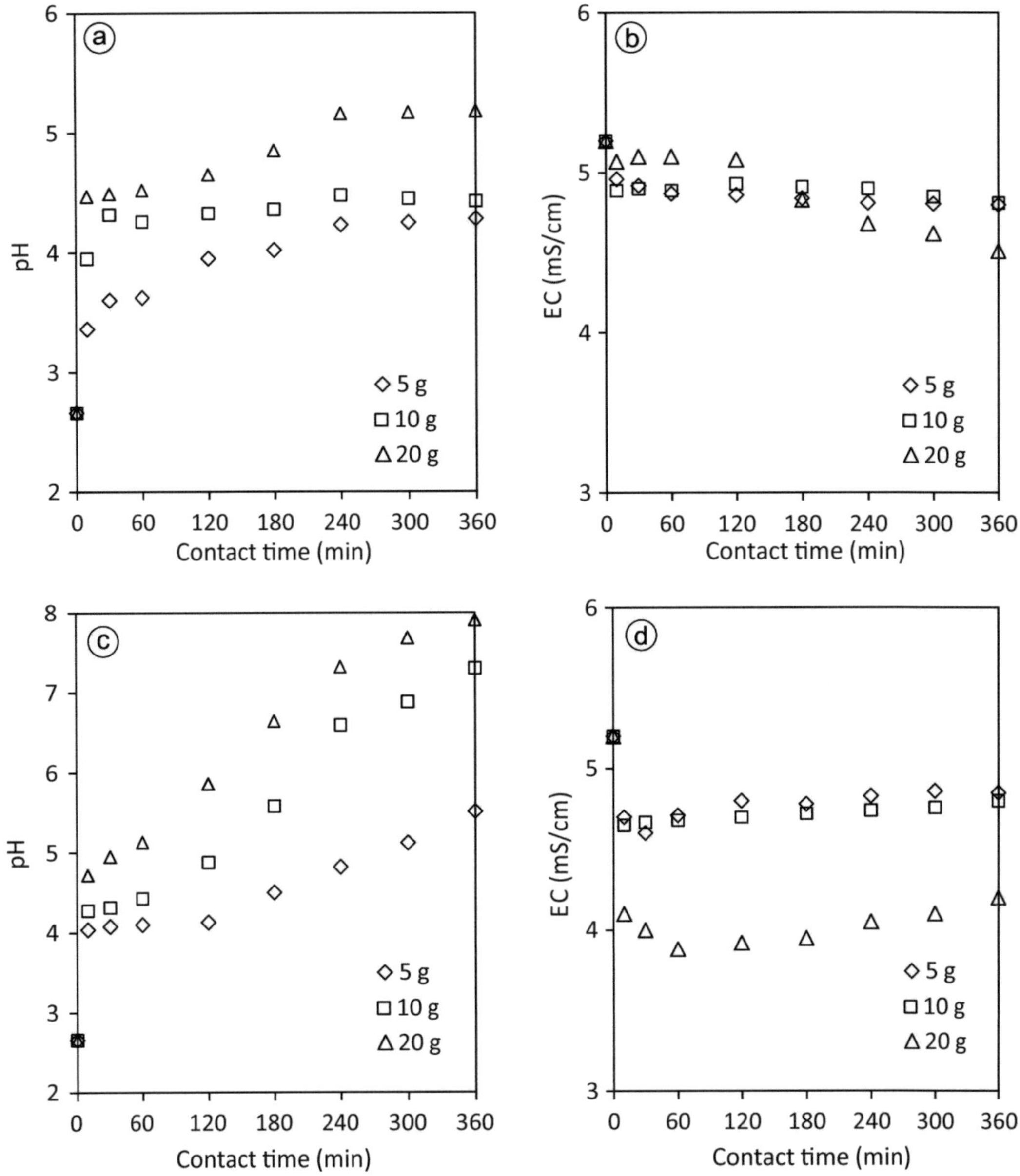

Figure 7. pH and EC vs. contact time as a function of adsorbent dosage, (a, b) FA, (c, d). MFA ($pH_{initial}$ = 2.66; $EC_{initial}$ = 5.2 mS/cm).

Fe concentration was determined 0.5 mg/L at pH = 5.51, 0.1 mg/L at pH = 7.3 and 0.05 mg/L at pH = 7.9 at the end of 6 hours, respectively (**Figure 8b**). It was observed that the adsorbent dosage is an effective parameter for Fe removal using FA and MFA into AMD. Previous studies reported that higher adsorption capacity and larger surface area of FA is related to

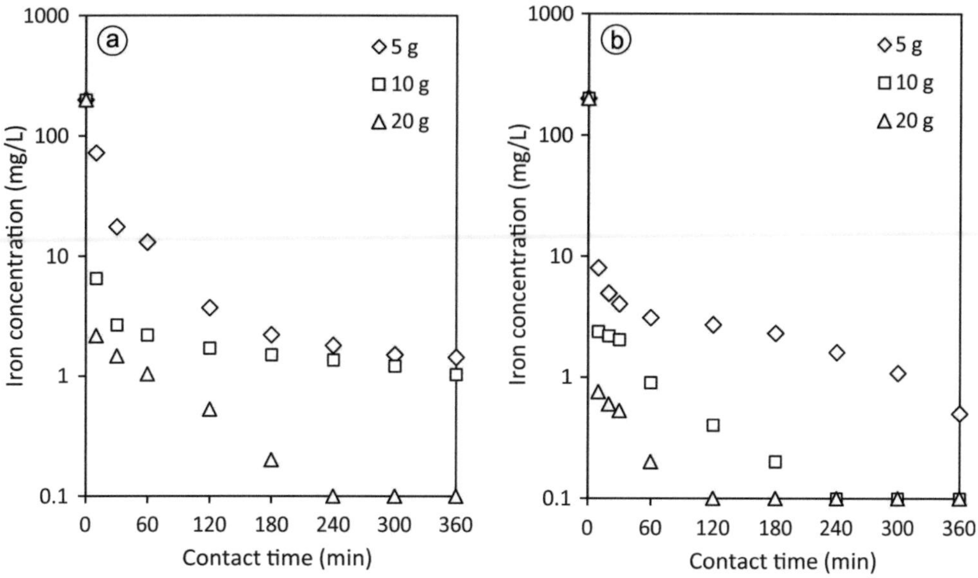

Figure 8. Effect of adsorbent dosage on the Fe concentration vs. contact time (a) FA and (b) MFA.

the efficiency of pollutant removal [54–58]. In this chapter, attempt was made to improve the efficiency of metal removal from AMD with modified FA using US mechanical effect. Modification of FA was provided to increase surface area and decrease diameter range of adsorbent and so, adsorbent dosage reduced by half using MFA. Iron concentration shows approximately 99.9% removal at pH = 7.3 attained by the MFA:AMD ratio of 1:10.

3.3.3. Effect of contact time

In order to determine the effect of contact time on the removal of Al, Fe and SO_4^{2-}, AMD (100 mL) was homogeneously mixed with FA (20 g) and MFA (10 g) at 180 rpm for 0–6 hours. **Figure 9a** and **b** shows the effect of contact time on Al, Fe and SO_4^{2-} removal from AMD. The pH of the solution (FA:AMD) ranged between 2.66 and 5.18. It is shown from **Figure 9a** that the removal percentage of Iron from solution increased rapidly for 20 g FA dosage and reached up to 99% at the end of 2 hours. A further increase in the contact time has a negligible effect on the rate of adsorption of Fe. Therefore, the optimum contact time of 2 hour for 20 g FA dosage could be considered for the adsorption of Fe on FA for entire batch studies. The pH of the solution (MFA:AMD) ranged between 2.66 and 7.3. The adsorb Fe on MFA increased with an increase of contact time in particular of the first 1 hour with 99% removal percentage. Iron concentration was successfully reduced below 0.3 mg/L at the end of 1 hour. Therefore, the optimum contact time of 1 hour for dosage of 10 g MFA could be considered for the adsorption of Fe on MFA for entire batch studies. Aluminium precipitates from AMD when they are diluted or neutralized to a pH of ≥ 4.5–5 [59] because pK1 = 5 for Al^{3+} hydrolysis, a necessary condition for onset of precipitation [60]. Removal percentage of Al from solution was calculated 85% (pH = 5.18) for dosage of 20 g FA at the end of 6 hours. Aluminium removal from solution was obtained as 86%

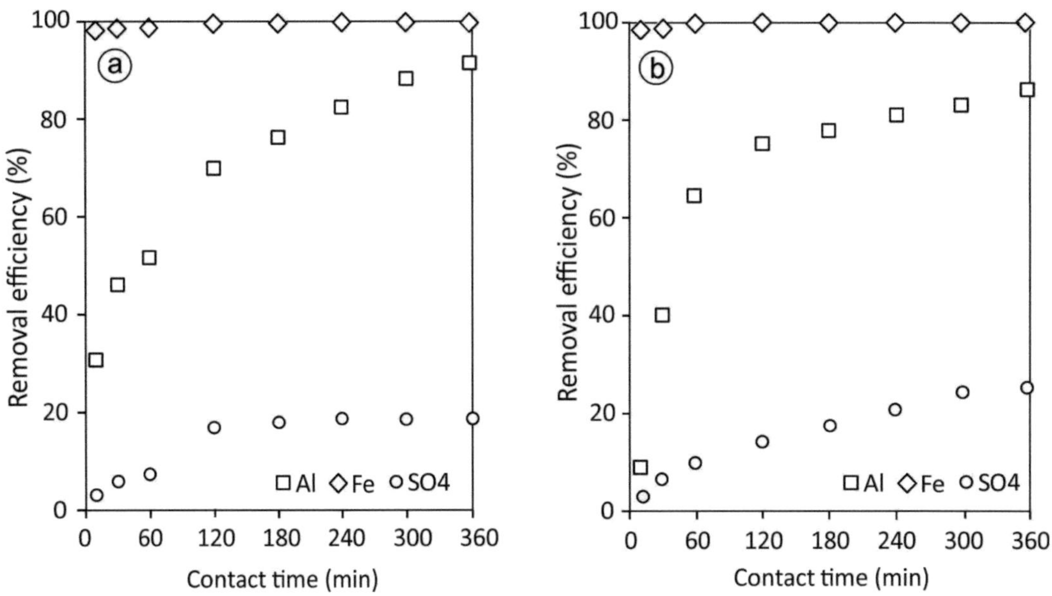

Figure 9. Effect of contact time on the removal of Al, Fe and SO_4^{2-} from AMD using (a) FA (20 g), (b) MFA (10 g).

(pH = 7.3) using 10 g MFA dosage at the end of 6 hours. Modification of FA was provided to reduce adsorbent dosage to half. Initial SO_4^{2-} concentration of AMD was measured 3600 mg/L. The maximum removal efficiency of SO_4^{2-} was obtained 18 and 25% for FA and MFA, respectively. According to the obtained results, the effectiveness of SO_4^{2-} removal from solution using FA and MFA by US enhanced with increasing ash dosage and contact time.

3.3.4. *Effect of pH*

pH is a major control on the solubility of most metal compounds [61]. The dependence of metal adsorption on pH is different for each metal [62]. Effect of various $pH_{initial}$ (3, 5, 7 and 9) on removal of selected metals (Al, Fe, Mn, Ni and Zn) and SO_4^{2-} from AMD were determined using 20 g dosage of FA and MFA for 6 hours contact time. The removal efficiency of Al, Fe, Mn, Ni and Zn was obtained 85.5, 99.9, 23.4, 67.5 and 51.2% for FA and 90.7, 99.9, 71.6, 93.8 and 99.4% for MFA at pH = 3, respectively (**Figure 10a** and **b**). The removal efficiency of Al, Fe, Mn, Ni and Zn was obtained 98.3, 99.9, 49.3, 83.7 and 99.3% for FA and 99.9, 99.9, 88.4, 97.8 and 99.7% for MFA at pH = 5, respectively. Ultrasound process can also accelerate chemical reactions [38]. The removal efficiency of metals increased in pH (7 and 9) reached over 97% removal. The obtained results showed that metal removal from AMD using FA and MFA depends on the final pH of solution. Sulphate removal as a function of different solutions and pH is shown in **Figure 10**. Approximately 39% of SO_4^{2-} can be removed from the solution (MFA:AMD) when the pH increased from 2.66 to 8.03.

3.4. Discussion

The AMD in Etili coal mine is highly acidic (pH of 2.66) because of intensive sulphide oxidation and ineffective neutralization processes. The abundance of pyrite in the volcanic rocks,

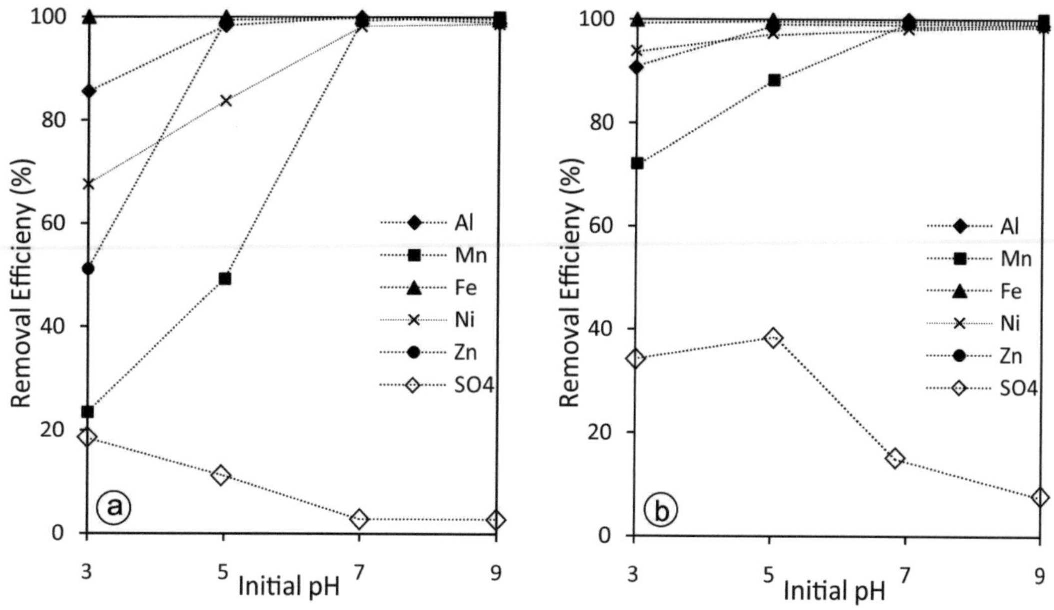

Figure 10. Effect of pH on removal efficiency at various metal and SO_4^{2-} from AMD using (a) FA (20 g) (b) MFA (20 g).

coal, mine wastes and sediments of mine lakes, as evidenced by XRD, SEM/EDX and isotope analyses of SO_4^{2-}, indicates that this mineral is the main sulphide mineral undergoing oxidation [3]. Sulphate is the predominant anion and Fe and Al are the most predominant metal species in AMD. Two anthropogenic pollutants (i.e. AMD and FA), usually formed in close proximity to each other, were reacted together in this chapter. Fly ash collected from the thermal power plant was modified with US process with the aim of activating surface area and decreasing particle size. The pH value of the ash-water system depends mainly on the Ca/S molar ratio in FA, although other minor alkalis or alkaline earth cations may also contribute to the balance [51, 63, 64]. Based on the Ca/S ratio and pH value, FA can be classified into three main groups: strongly alkaline ash (pH 11–13), mildly alkaline ash (pH 8–9) and acidic ash [51]. Fly ash is strongly alkaline, and its surface is negatively charged at high pH (see **Table 2**). Hence, it can be expected to remove metal ions from solutions by precipitation or electrostatic adsorption [62]. Besides, it contains a certain volume of unburnt carbon, which has a high adsorption capacity. The major inorganic species Al, Fe, Mn and SO_4^{2-} seem to be mainly removed through precipitation of amorphous oxides, (oxy)hydroxides and basic hydroxyl sulphates [16]. The AMD was characterized by high concentrations of Fe^{3+} as compared to Fe^{2+} (**Table 1**). The initial decrease in concentration for both ratios at pH 4–4.5 indicates removal of Fe^{3+} in the form of hydroxides or (oxy)hydroxysulphates [16]. Aluminium and Fe mobility at mining sites is related to pH and SO_4^{2-} concentration [65]. Saturation of iron-hydroxysulphate minerals generally occurs around $pH = 4$ and so, at $pH > 4$, Fe precipitates as a solid and is no longer mobile [61, 65]. Dissolution of FA and MFA components is confirmed through the decrease of Fe in the Fe-oxide fraction at pH 4.5–4.7. It is often suggested that the tendency of metal cations to adsorb on the oxide surfaces is highly related to their pH of hydrolysis [62, 66]. At pH greater than 7, 90.7% of the Al was removed from solution at 20 g MFA dosage and significant proportion of Al from AMD was hydrolysed to form amorphous (oxy)hydroxides. It was observed that the maximum removal efficiency for Zn and Ni was achieved using FA and MFA at pH of 8. The leaching of Ca and Mg can be assumed to be controlled by the dissolution of Ca and Mg-bearing silicates [67, 68]. According to ICP-MS results, Ca, Mg, Na and K concentrations were observed to increase from 496 to 645 and 802 mg/L; 267 to 335 and 311 mg/L; 156 to 221 and 166 mg/L; 3 to 36 and 16 mg/L for FA:AMD and MFA:AMD, respectively. Calcium, Mg, Na and K concentrations decreased in SR 0.2, 0.5, 0.11 and 0.02% for FA:AMD; 0.22, 0.14, 0.02 and 0.07% for MFA:AMD, respectively, in XRF analysis results (**Table 3**). The removal of SO_4^{2-} depends on CaO containing FA. Soluble CaO (1.74–1.81%) in FA and MFA increased the pH value of AMD. According to particle size analysis, the particle size distribution of SR (FA:AMD) increased 31, 18 and 12% for Dx (10, 50 and 90), respectively. The particle size distribution of SR (MFA:AMD) increased 5, 6 and 12% for Dx (10, 50 and 90), respectively (**Figure 11**). Fly ash and MFA reacted with AMD were explained based on the particle morphology, shape and mineralogical components. These interpretations from the particle surface analysis were supported with SEM images. **Figure 12a** and **b** shows the SEM images of SR (MFA:AMD). Modified fly ash particles viewed in a predominantly irregular-shaped and have pores and creeks. These results confirm that the surfaces of FA particles were disrupted with US shear forces. According to EDX analysis results, Al, Fe, Mn and Zn concentrations in SR accumulated and precipitated on adsorbent surface. The XRD analysis of the sonicated and unsonicated FA showed no mineral phase change (**Table 4**). Gypsum was the only new mineral phase after treatment of AMD with FA and MFA was identified by XRD analysis in the SR. Gypsum was also detected by SEM-EDX analysis (**Figure 12c**). Gypsum crystals are containing 83.1% O, 8% S and 6.7% Ca, as quantified by EDX analysis (**Figure 12d**). As the final pH of solution increased, more CaO is available and more Ca^{2+} ions are resulting in enhanced removal of SO_4^{2-} as gypsum. The FTIR spectra of the FA, MFA and SR are illustrated in **Figure 13**. The FTIR data validate the observations made through XRD. The main band, due

Major component (wt%)	**FA:AMD**	**MFA:AMD**
SiO_2	49.5	49.28
Al_2O_3	20.17	19.99
MgO	5.28	5.36
Fe_2O_3	5.60	5.67
Na_2O	2.01	2.03
CaO	1.68	1.45
K_2O	1.70	1.64
MnO	0.072	0.07
ZnO	0.018	0.019
NiO	0.008	0.008
CuO	0.007	0.007
S	0.43	0.40
LOI	11.92	13.41

Table 3. Chemical composition of SR.

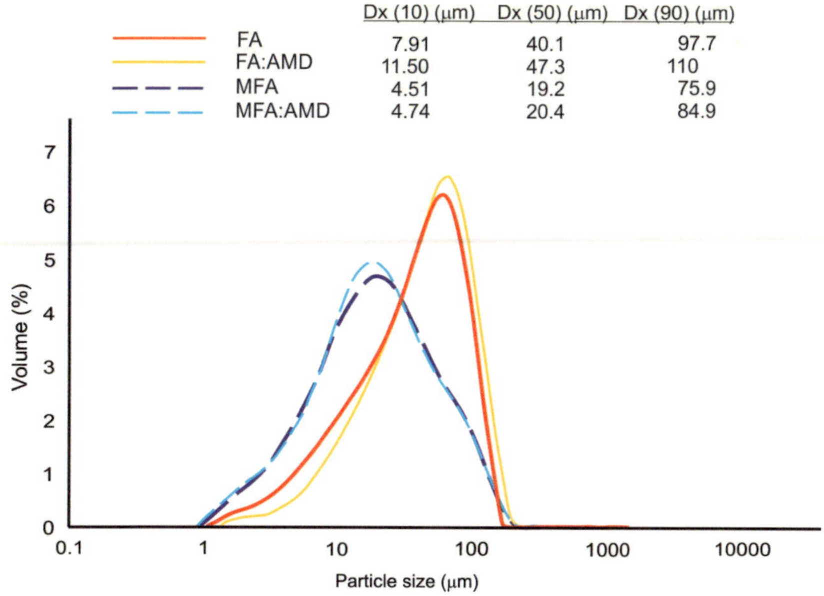

Figure 11. Particle size distribution of FA, MFA and SR.

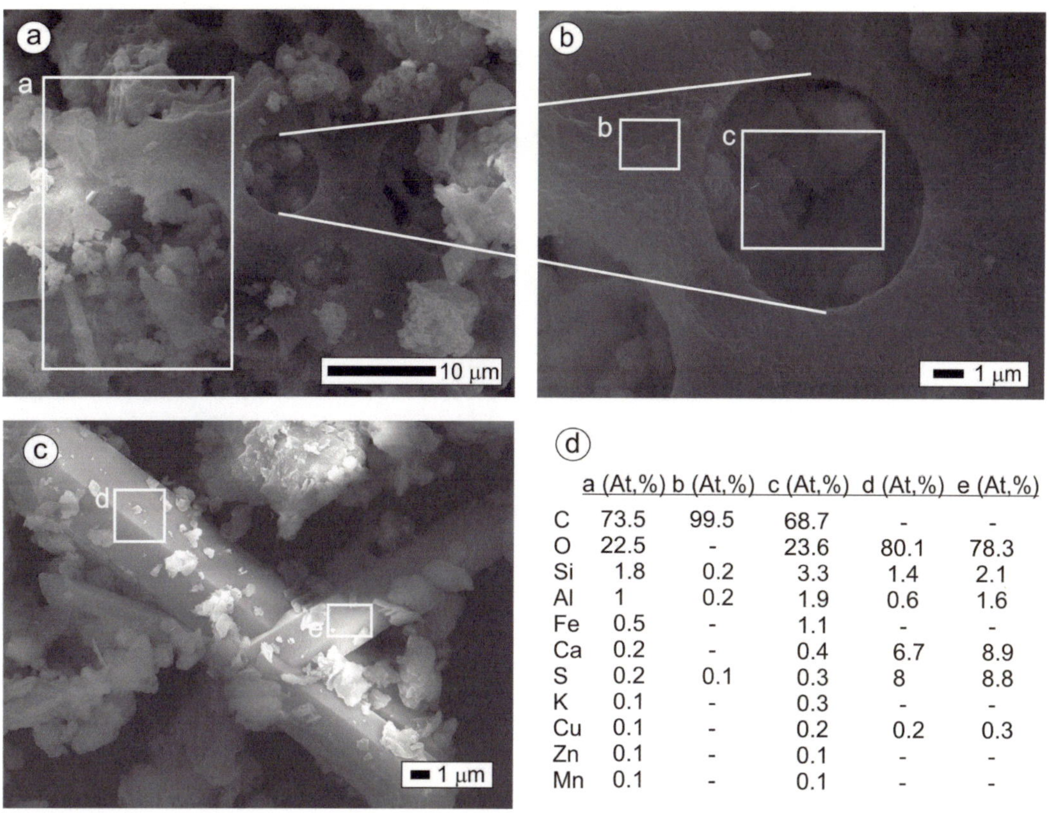

	a (At,%)	b (At,%)	c (At,%)	d (At,%)	e (At,%)
C	73.5	99.5	68.7	-	-
O	22.5	-	23.6	80.1	78.3
Si	1.8	0.2	3.3	1.4	2.1
Al	1	0.2	1.9	0.6	1.6
Fe	0.5	-	1.1	-	-
Ca	0.2	-	0.4	6.7	8.9
S	0.2	0.1	0.3	8	8.8
K	0.1	-	0.3	-	-
Cu	0.1	-	0.2	0.2	0.3
Zn	0.1	-	0.1	-	-
Mn	0.1	-	0.1	-	-

Figure 12. SEM results of: (a, b) SR (MFA:AMD), (c) gypsum and (d) EDX results of SR.

Mineral phases	FA	MFA	FA:AMD	MFA:AMD
	%			
Quartz	42	37	40	33
Amorphous phase	23	31	28	29
Mullite	13	10	7	9
Plagioclase	12	10	8	9
Gypsum	-	-	8	12
Hematite	6	2	4	3
Magnetite	2	1	1	1
Mica	1	5	2	3
K-feldspar	1	4	2	1

Table 4. Mineral phases in FA, MFA and SR.

to Si-O and Al-O vibrations, which is located at 1046 cm^{-1} in FA, moves towards lower frequencies when those are activated [69]. Ultrasound energy strengthens the bond at the solid particle phase interface [38]. Band at about 1000 cm^{-1}, corresponding either to the stretching vibrations of Si-O bonds in SiO_4 or to the skeleton of bonded SiO_4 tetrahedra, shows the greatest change during crystallization [38, 70, 71]. The FTIR spectrum of the FA and MFA of crystallization contains a number of characteristic vibration bands in the regions 1100–900 and 850–650 cm^{-1}.

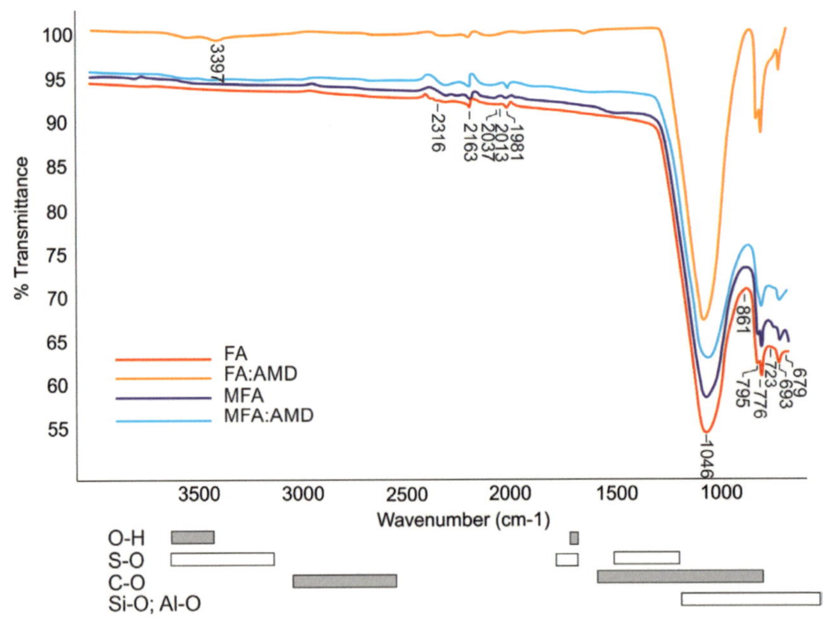

Figure 13. FTIR spectra of FA, MFA and SR.

4. Conclusion

Due to unique properties of the all mines, AMD may display variable hydrogeochemical characteristics; special studies are required for each mine site, though the general approach remains the same. In this chapter, treatment of AMD using FA and MFA under various conditions was investigated. The effect of US on the surface of FA was ensured adsorbent surface activation by ultrasonic cavitation. The physical changes on the surface of the FA caused by the US-assisted modification were identified using arrays of XRF, BET, SEM-EDX, XRD, FTIR techniques and particle size analysis. It was observed that due to the ultrasound-assisted modification, beside the physical properties of FA, the morphological and mineralogical characteristics are varied as well. The MFA has a lower Si and Al concentrations and higher surface area and pH than the FA. The chemical results imply that it is possible to use FA and MFA in order to neutralize AMD followed by precipitation and adsorption of metals. The 99.9% Fe removal was determined using FA with optimum 2 hours contact time and 1:5 (FA:AMD) ratio and using MFA with optimum 1 hour contact time and 1:10 (MFA:AMD) ratio under the same experimental conditions. Modified fly ash provides more successful results in a shorter reaction time and with smaller FA dosage for metal and SO_4^{2-} removal from AMD when compared to FA. Acid mine drainage reacted with MFA resulted in neutral pH after a longer reaction time was accompanied by 39% SO_4^{2-} removal. The five factors that dictate the nature of the final solution in neutralization reactions and metal removal are modification, mixing speed, pH, adsorbent dosage and contact time. Preliminary treatment of AMD revealed that the MFA has the potential to be used as an effective adsorbent to treat AMD. Additionally, this chapter provides an example for other areas with AMD environmental problems.

Acknowledgements

The described work was carried out as part of FBA-2017-1077 numbered project. This project is supported by the Canakkale Onsekiz Mart University, cofinanced by the Coordinatorship of Scientific Research. The authors are grateful to Editors for their constructive comments.

Author details

Deniz Sanliyuksel Yucel[1]* and Burcu Ileri[2]

*Address all correspondence to: denizsyuksel@comu.edu.tr

1 Department of Geological Engineering, Canakkale Onsekiz Mart University, Canakkale, Turkey

2 Department of Environmental Engineering, Canakkale Onsekiz Mart University, Canakkale, Turkey

References

[1] Gurdal G, Bozcu M. Petrographic characteristics and depositional environment of Miocene Can coals, Canakkale-Turkey. The International Journal of Coal Geology. 2011;**85**(1):143-160. DOI: 10.1016/j.coal.2010.11.001

[2] Sanliyuksel Yucel D, Yucel MA, Baba A. Change detection and visualization of acid mine lakes using time series satellite image data in geographic information systems (GIS): Can (Canakkale) County, NW Turkey. Environmental Earth Sciences. 2014;**72**(11):4311-4323. DOI: 10.1007/s00244-016-0270-z

[3] Sanliyuksel Yucel D, Balci N, Baba A. Generation of acid mine lakes associated with abandoned coal mines in northwest Turkey. Archives of Environmental Contamination and Toxicology. 2016;**70**(4):757-782. DOI: 10.1007/s00244-016-0270-z

[4] Yucel MA, Turan RY. Areal change detection and 3D modeling of mine lakes using high-resolution unmanned aerial vehicle images. Arabian Journal for Science and Engineering. 2016;**41**(12):4867-4878. DOI: 10.1007/s13369-016-2182-7

[5] Sanliyuksel Yucel D, Baba A. Geochemical characterization of acid mine lakes and their effect on the environment, NW of Turkey. Archives of Environmental Contamination and Toxicology. 2013;**64**(3):357-376. DOI: 10.1007/s00244-012-9843-7

[6] Feng D, Aldrich C, Tan H. Treatment of acid mine water by use of heavy metal precipitation and ion exchange. Minerals Engineering. 2000;**13**(6):623-642. DOI:10.1016/S0892-6875(00)00045-5

[7] Ulubas T. Use of fly ash for heavy metal (Cu^{+2}, Pb^{+2}) removal [thesis]. Istanbul: Istanbul Technical University; 2000

[8] Banerjee SS, Jayaram RV, Joshi MV. Removal of nickel(II) and zinc(II) from wastewater using fly ash and impregnated fly ash. Separation Science and Technology. 2006;**38**(5):1015-1032. DOI: 10.1081/SS-120018121

[9] Reynolds K. The use of fly ash for the control and treatment of acid mine drainage. In: Proceedings of the World of Coal Ash (WOCA); 11-15 April 2005; Johannesburg. Lexington: WOCA; 2005. p. 956-961

[10] Skousen JG, Sexstone A, Ziemkiewicz PF. Acid mine drainage control and treatment. 1st ed. The American Society of Agronomy and Academic Press; Reclamation of Drastically Disturbed Lands (Agronomy, No. 41); 2000. pp. 105-129. DOI:10.2134/agronmonogr41.c5

[11] Rios CA, Williams CD, Roberts CL. Removal of heavy metals from acid mine drainage (AMD) using coal fly ash, natural clinker and synthetic zeolites. Journal of Hazardous Materials. 2008;**156**(1-3):23-35. DOI: 10.1016/j.jhazmat.2007.11.123

[12] Ahmaruzzaman MA. Review on the utilization of fly ash. Progress in Energy and Combustion Science. 2010;**36**(3):327-336. DOI: 10.1016/j.pecs.2009.11.003

[13] Abas SNA, Ismail MHS, Kamala ML, Izhar S. Adsorption process of heavy metals by low-cost adsorbent: A review. World Applied Sciences Journal. 2013;**28**(11):1518-1530. DOI: 10.5829/idosi.wasj.2013.28.11.1874

[14] Stouraiti C, Xenidis A, Paspaliaris I. Reduction of Pb, Zn and Cd availability from tailings and contaminated soils by the application of lignite fly ash. Water Air and Soil Pollution. 2001;**137**(1-4):247-265. DOI: 10.1023/A:1015507017799

[15] Okafor EC, Opuene K. Preliminary assessment of trace metals and polycyclic aromatic hydrocarbons in the sediments. International Journal of Environmental Science and Technology. 2007;**4**(2):233-240. DOI: 10.1007/BF03326279

[16] Gitari WM, Petrik LF, Key DL, Okujeni C. Partitioning of major and trace inorganic contaminants in fly ash acid mine drainage derived solid residues. International Journal of Environmental Science and Technology. 2010;**7**(3):519-534. DOI: 10.1007/BF03326161

[17] Tutunlu F, Atalay U. Utilization of fly ash in manufacturing of building brick. In: Proceedings of the International Ash Utilization Symposium, Center for Applied Energy Research, University of Kentucky; 22-24 October 2001; Ankara. Lexington: CAER; 2001. pp. 13-18

[18] Ozdemir O, Celik MS. Characterization and recovery of lignitic fly ash by products from the Tuncbilek Power Station. Canadian Metallurgical Quarterly. 2002;**41**(2):143-150. DOI: 10.5277/ppmp160206

[19] Ozdemir O, Ersoy B, Celik MS. Separation of pozzolonic material from lignitic fly ash of Tuncbilek Power Station. In: Proceedings of the International Ash Utilization Symposium, Center for Applied Energy Research, University of Kentucky; 22-24 October 2001; Istanbul. Lexington: CAER; 2001. pp. 45-52

[20] Celik O, Damcı E, Piskin S. Characterization of fly ash and its effects on the compressive strength properties of Portland Cement. Indian Journal of Engineering and Materials Sciences. 2008;**15**(5):433-440

[21] Ersoy B, Kavas T, Evcin A, Baspinar S, Sariisik A, Dikmen S. Production of fired ceramic materials from fly ash with witherite additive. Afyon Kocatepe University Journal of Science. 2009;**9**(3):45-52

[22] Gitari MW, Petrik LF, Etchebers O, Key DL, Iwuoha E, Okujeni C. Treatment of acid mine drainage with fly ash: Removal of major contaminants and trace elements. Journal of Environmental Science and Health Part A. 2006;**41**(8):729-1747. DOI: 10.1080/10934520600754425

[23] Pérez-López R, Nieto JM, Almodóvar GR. Utilization of fly ash to improve the quality of the acid mine drainage generated by oxidation of a sulphide-rich mining waste: Column experiments. Chemosphere. 2007;**67**(8):1637-1646. DOI:10.1016/j.chemosphere.2006.10.009

[24] Gitari MW, Petrik LF, Etchebers O, Key DL, Iwuoha E, Okujeni C. Passive neutralization of acid mine drainage by fly ash and its derivatives: A column leaching study. Fuel. 2008;**87**(8-9):1637-1650. DOI: 10.1016/j.fuel.2007.08.025

[25] Gitari WM, Petrik LF, Etchebers O, Key DL, Okujeni C. Utilization of fly ash for treatment of coal mines wastewater: Solubility controls on major inorganic contaminants. Fuel. 2008;**87**(12):2450-2462. DOI: 10.1016/j.fuel.2008.03.018

[26] Vadapalli VRK, Klink MJ, Etchebers O, Petrik LF, Gitari W, White RA, Key D, Iwuoha E. Neutralization of acid mine drainage using fly ash, and strength development of the resulting solid residues. South African Journal of Science. 2008;**104**(7-8):317-322. ISSN: 1996-7489

[27] Sahoo PK, Tripathy S, Panigrahi MK, Equeenuddin SM. Evaluation of the use of an alkali modified fly ash as a potential adsorbent for the removal of metals from acid mine drainage. Applied Water Science. 2013;**3**(3):567-576. DOI:10.3390/met6120300

[28] Qureshi A, Jia Y, Maurice C, Öhlander B. Potential of fly ash for neutralisation of acid mine drainage. Environmental Science and Pollution Research. 2016;**23**(17):17083-17094. DOI: 10.1007/s11356-016-6862-3

[29] Mason TJ, Peters D. Pratical Sonochemistry, Power Ultrasound Uses and Applications. 2nd ed. Woodgate: Horwood Publishing Limited; 2002. p. 155. ISBN: 978-1898563839

[30] Brotchie A, Borisova D, Belova V, Möhwald H, Shchukin D. Ultrasonic modification of aluminum surfaces: Comparison between thermal and ultrasonics effects. Journal of Physical Chemistry. 2012;**116**(14):7952-7956. DOI: 10.1021/jp3016408

[31] Geiger CL, Ruiz NE, Clausen CA, Reinhart DR, Quinn JW. Ultrasound pretreatment of elemental iron: Kinetic studies of dehalogenation reaction enhancement and surface effects. Water Research. 2002;**36**(5):1342-1350. DOI: 10.1016/S0043-1354(01)00319-0

[32] Morais NLPA, Brett CMA. Influence of ultrasound on the corrosion of aluminum. Key Engineering Material. 2002;**230-232**:412-415. DOI: 10.4028/www.scientific.net/KEM.230-232.412

[33] Liang F, Fan J, Guo Y, Fan M, Wang J, Yan H. Reduction of nitrite by ultrasound-dispersed nanoscale zero-valent iron particle. Industrial and Engineering Chemistry Research. 2008;**47**(22):8550-8554. DOI: 10.1021/ie8003946

[34] Adewuyi YG. Sonochemistry: Environmental science and engineering applications. Industrial&Engineering Chemistry Research. 2001;**40**(22):4681-4715. DOI: 10.1021/ie0100961

[35] Adewuyi YG. Sonochemistry in environmental remediation. 1. Combinative and hybrid sonophotochemical oxidation processes for the treatment of pollutants. Water Environmental Science & Technology. 2005;**39**(10):3409-3420. DOI: 10.1021/es049138y

[36] Villeneuve L, Alberti L, Steghens JP, Lancelin JM, Mestas JL. Assay of hydroxyl radicals generated by focused ultrasound. Ultrasonics Sonochemistry. 2009;**16**(3):339-344. DOI: 10.1016/j.ultsonch.2008.09.007

[37] Ileri B, Ayyildiz O, Apaydin O. Ultrasound-assisted activation of zero-valent magnesium for nitrate denitrification: Identification of reaction by-products and pathways. Journal of Hazardous Materials. 2015;**292**:1-8. DOI: 10.1016/j.jhazmat.2015.03.004

[38] Bukhari SS. Microwave and ultrasound assisted zeolitization of coal fly ash [thesis]. Ontario: Western University; 2016

[39] Bukhari SS, Rohani S, Kazemian H. Effect of ultrasound energy on the zeolitization of chemical extracts from fused coal fly ash. Ultrasonics Sonochemistry. 2016;**28**:47-53. DOI: 10.1016/j.ultsonch.2015.06.031

[40] Kalipci E, Sahinkaya S, Aras S, Ozturk M. Chromium (VI) adsorption with ultrasonically activated carbon. Journal of Nevsehir University Institute of Science. 2013;**2**(1):20-25

[41] Chapman D, Kimstach V. Selection of water quality variables. In: Chapman D, editor. Water Quality Assessments-A Guide to Use of Biota, Sediments and Water in Environmental Monitoring. 2nd ed. E & FN Spon: Chapmam & Hall; 1996. pp. 59-126. ISBN: 978-0419216001

[42] Okumusoglu D, Gunduz O. Hydrochemical status of an acidic mining lake in Can-Canakkale, Turkey. Water Environment Research. 2013;**85**(7):604-620. DOI: 10.2175/106143013X13698672321463

[43] Kirby CS, Cravotta III CA. Net alkalinity and net acidity 1: Theoretical considerations. Applied Geochemistry. 2005;**20**(10):1920-1940. DOI:10.1016/j.apgeochem.2005.07.002

[44] SKKY. Turkish Water Pollution Control Regulation, (TWPCR), 2008. The Regulation of Water Pollution Control. Ministry of Environment and Forestry. Official Newspaper. 31.12.2004:Number: 25687

[45] Shreya N, Paul B. Utilization of fly ash as a carrier in biofertilizer and biopesticide formulation of chandrapura thermal power station, India. In: Proceedings of Twenty-Seventh International Conference on Solid Waste Technology and Management (ISCW-2012); 11-14 March 2012; Jharkhand. Pennsylvania: ISCW; 2012. p. 24

[46] Sanliyuksel Yucel D. Removal of heavy metals from aqueous solution using fly ash: Can Thermal Power Plant, NW Turkey as a Case Study. Karaelmas Science and Engineering Journal. 2017;**7**(1):291-298. DOI: http://dx.doi.org/10.7212%2Fzkufbd.v1i1.288

[47] ASTM. Standard C618-15 Standard Specification for Coal Fly Ash and Raw or Calcined Natural Pozzolan for use in Concrete. West Conshohocken, PA, USA: ASTM International; 2015

[48] Lo TY, Cui H, Memon SA, Noguchi T. Manufacturing of sintered lightweight aggregate using high-carbon fly ash and its effect on the mechanical properties and microstructure of concrete. Journal of Cleaner Production. 2016;**112**(1):753-762. DOI: 10.1016/j.jclepro.2015.07.001

[49] Lee GS, Han GY, Kim SD. Coal combustion characteristics in a circulating fast fluidized bed. Korean Journal of Chemical Engineering. 1984;**1**(1):71-76. DOI: 10.1007/BF02697421

[50] Seo MW, Goo JH, Kim SD, Lee SH, Choi YH. Gasification characteristics of coal/biomass blend in a dual circulating fluidized bed reactor. Energy Fuel. 2010;**24**(5):3108-3118. DOI: 10.1021/ef100204s

[51] Yao ZT, Ji XS, Sarker PK, Tang JH, Ge LQ, Xia MS, Xi YQ. A comprehensive review on the applications of coal fly ash. Earth-Science Reviews. 2015;**141**:105-121. DOI: 10.1016/j.earscirev.2014.11.016

[52] Ragheb SM. Phosphate removal from aqueous solution using slag and fly ash. HBRC Journal. 2013;**9**(3):270-275. DOI: 10.1016/j.hbrcj.2013.08.005

[53] Ibis RY. Investigation of the efficacy of the thermal power plant fly ash removal of arsenic from waste water [thesis]. Adana: Cukurova University; 2010

[54] Bayat B. Comparative study of adsorption properties of Turkish fly ashes I. The case of nickel(II), copper(II) and zinc(II). Journal of Hazardous Materials. 2002;**95**(3):251-273. DOI: 10.1016/S0304-3894(02)00140-1

[55] Bayat B. Comparative study of adsorption properties of Turkish fly ashes II. The case of chromium (VI) and cadmium (II). Journal of Hazardous Materials. 2002;**95**(3):275-290. DOI: 10.1016/S0304-3894(02)00141-3

[56] Mohan S, Gandhimathi R. Removal of heavy metal ions from municipal solid waste leachate using coal fly ash as an adsorbent. Journal of Hazardous Materials. 2009;**169**(1-3):351-359. DOI: 10.1016/j.jhazmat.2009.03.104

[57] Shah AK, Zeenat MA, Memon AR, Laghari AJ, Mughal MA, Shah SFA, Saleem H. Exploitation of low cost coal fly ash adsorbent with coagulants for the treatment of industrial complex nature dyes wastewater. International Journal of Scientific & Engineering Research. 2013;**4**(9):109-119. ISSN: 2229-5518

[58] Kulkarni SJ, Dhokpande SR, Kaware JP. Studies on fly ash as an adsorbent for removal of various pollutants from wastewater. International Journal of Engineering Research& Technology. 2013;**2**(5):1190-1195. ISSN: 2278-0181

[59] Nordstrom DK, Ball JW. The geochemical behavior of aluminum in acidified surface waters. Science. 1986;**232**(4746):54-56. DOI: 10.1126/science.232.4746.54

[60] Nordstrom DK. Hydrogeochemical processes governing the origin, transport and fate of major and trace elements from mine wastes and mineralized rock to surface waters. Applied Geochemistry. 2011;**26**(11):1777-1791. DOI: 10.1016/j.apgeochem.2011.06.002

[61] Smith KS. Strategies to predict metal mobility in surficial mining environments. The Geological Society of America Reviews in Engineering Geology. 2007;**17**:25-45. DOI: 10.1130/2007.4017(03)

[62] Cho H, Oh D, Kim K. A study on removal characteristics of heavy metals from aqueous solution by fly ash. Journal of Hazardous Materials. 2005;**127**(1-3):187-195. DOI: 10.1016/j.jhazmat.2005.07.019

[63] Ward CR, French D, Jankowski J, Dubikova M, Li Z, Riley KW. Element mobility from fresh and long-stored acidic fly ashes associated with an Australian power station. International Journal Coal Geology. 2009;**80**(3-4):224-236. DOI: 10.1016/j.coal.2009.09.001

[64] Izquierdoa M, Querol X. Leaching behaviour of elements from coal combustion fly ash: An overview. International Journal Coal Geology. 2012;**94**:54-56. DOI: 10.1016/j.coal.2011.10.006

[65] Degraff JV. Addressing the toxic legacy of abandoned mines on public land in the western United States. In: Degraff JV, editor. Understanding and Responding to Hazardous Substances at Mine Sites in the Western United States, Reviews in Engineering Geology XVII. 1st ed. Boulder: The Geological Society of America; 2007. pp. 1-8. DOI: 10.1130/2007.4017(01)

[66] Gitari WM, Fatoba OO, Petrik LF, Vadapalli RKV. Leaching characteristics of selected South African fly ashes: Effect of pH on the release of major and trace species. Journal of Environmental Science and Health Part A. 2009;**44**(2):206-220. DOI: 10.1080/10934520802539897

[67] Rigol A, Mateu J, Gonzalez-Nunez R, Rauret G, Vidal M. pH stat vs. single extraction tests to evaluate heavy metals and arsenic leachability in environmental samples. Analytica Chimica Acta. 2009;**632**(1):69-79. DOI: 10.1016/j.aca.2008.10.066

[68] Silva LFO, Izquierdo M, Querol X, Finkelman RB, Oliveira MLS, Wollenshlager M, Towler M, Pérez-López R, Macias F. Leaching of potential hazardous elements of coal cleaning rejects. Environmental Monitoring and Assessment. 2011;**175**(1):109-126. DOI: 10.1007/s10661-010-1497-1

[69] Palomoa A, Grutzeck MW, Blanco MT. Alkali-activated fly ashes a cement for the future. Cement and Concrete Research. 1999;**29**(8):1323-1329. DOI: 10.1016/S0008-8846(98)00243-9

[70] Stojkovic SR, Adnadjevic B. Investigation of the NaA zeolite crystallization mechanism by i.r. spectroscopy. Zeolites. 1988;**8**(6):523-525. DOI: 10.1016/S0144-2449(88)80230-6

[71] De Man AJM, Van Santen RA. The relation between zeolite framework structure and vibrational spectra. Zeolites. 1992;**12**(3):269-279. DOI: 10.1016/S0144-2449(05)80295-7

Chapter 5

Treatment of Acid Mine Drainage with Coal Fly Ash: Exploring the Solution Chemistry and Product Water Quality

Wilson Mugera Gitari, Leslie F. Petrik and
Segun A. Akinyemi

Additional information is available at the end of the chapter

http://dx.doi.org/10.5772/intechopen.69741

Abstract

A treatment process for Acid mine drainage (AMD) using coal fly ash (CFA) was developed. AMD was treated with CFA as the alkaline agent at different CFA: AMD ratios and pH, electrical conductivity (EC) evolution monitored over time. In a separate experiment two AMD sources with differing chemistry were treated with the same CFA to evaluate the impact of AMD chemistry on the treatment process and product water quality. Various CFA: AMD ratios were stirred in a beaker for a pre-set time and the process water chemistry determined. pH was observed to increase in a stepwise manner with buffer zones observed at 4-4.5, 4.5-7 and 6-8. AMD with low concentration of Al^{3+}, Fe^{2+}, Fe^{3+} and Mn^{2+} didn't exhibit these buffer zones. Two competing processes were observed to control the evolving pH of process water: dissolution of basic oxides (CaO, MgO) from CFA led to pH increase and hydrolysis of AMD species such as Al^{3+}, Fe^{2+}, Fe^{3+} and Mn^{2+} led to pH decrease. These processes initiated mechanisms such as precipitation, adsorption and ion exchange that led to decrease in inorganic contaminants as the treatment progressed. Inorganic contaminants removal was directly related to amount of CFA in reaction media. Precipitation of insoluble hydroxides and Al, Fe-oxyhydroxysulphates contributed to removal of major and minor chemical species. Precipitation of gypsum contributed to removal of sulphate. Na, K and Mg remained largely in solution after initial decrease. Significant leaching of B, Sr, Ba, and Mo from CFA was observed and was directly linked to amount of CFA in the reaction media. This will be a shortcoming of the treatment process since other processes may be required to polish up the product water. The treatment of AMD with CFA was observed to depend on CFA, AMD chemistry, treatment time and might therefore be site specific.

Keywords: acid mine drainage, coal fly ash, sulfate, neutralization, inorganic contaminants, pH, coal fly ash: acid mine drainage ratios

1. Introduction

The mining industry in South Africa has a huge potential to impact negatively on the environment. Negative impacts include generation of reactive tailings and acid mine drainage (AMD). In South Africa, the Witwatersrand basin alone decants approximately 10–60 mL/day into nearby rivers [1]. AMD is highly acidic (pH 2–4), sulfate-rich and frequently carries a high concentration of inorganic contaminants. AMD is extremely acidic (as low as pH 2.0) and enriched with iron, manganese, aluminum, sulfate, and metal species such as lead, mercury, cadmium, zinc [2–6]. When sulfide minerals such as pyrite (FeS_2), its dimorph marcasite and pyrrhotite ($Fe_{1-x}S$) are exposed to oxygen and water, they undergo a bacterially catalyzed oxidation reaction which lead to generation of acidity and increased Fe and sulfate concentrations in recipient water bodies (Eq. (1)). The overall reaction is often written as [7]:

$$FeS_2(s) + 3.75O_2 + 3.5H_2O \rightarrow Fe(OH)_3(s) + 2SO_4^{2-} + 4H^+ \quad (1)$$

On interacting with the mine bedrock, the acidic water leaches more chemical species leading to high concentrations of Fe, Mn, Al, Cu, Zn, Mg, Na and Ni in the AMD streams [2, 3].

South Africa environmental regulations require these AMD streams to be treated to acceptable levels before discharging into surface water bodies. A range of active and passive remediation technologies have been adopted by various mining companies to reduce impact of AMD on ground and surface water resources. These technologies include active neutralization by lime, limestone, biological sulfate removal [8–10], eutectic freeze crystallization [11] and the alkaline barium calcium desalination process (ABC) [12]. Passive AMD remediation technologies include alkalinity generating artificially constructed wetlands [13]. Other innovative AMD management and remediation technologies have been developed which involve treatment and recovery of beneficial products such as gypsum and sulfur [12]. Recently, innovative and sustainable management technologies for AMD have been developed which involve treatment and recovery of drinking water that is supplied to communities near the mines while the gypsum recovered is used in construction of gypsum boards and houses. This is to reduce the cost of treatment and make the process sustainable [14]. Bhattacharyya and Gupta [15] observe that these AMD management technologies are still expensive and mining companies are still searching for cheaper treatment and management technologies.

South Africa uses more than 100 MT of low grade bituminous coal annually to produce electricity in coal powered utilities. These coal power utilities in turn produce 28 MT of coal fly ash annually that is disposed of on land as ash dumps or slurried to ash dams due to lack of alternative application of this coal combustion by-product [16]. Fly ash is a powdery substance which consists of fine spherical particles which are either solid or hollow. It is a ferro-aluminosilicate material with particles ranging in size from 20 to 80 μm, its surface is enriched with trace elements of Si, Al, Fe, Ca, K and Na [17, 18]. Several authors have reported that South African coal fly ash consists mainly of aluminosilicate glass matrix in addition to crystalline mullite and quartz. This glass phase was observed to be associated with elements such as As, Na, Mg, K, Sr, B and Mo as soluble salts on the surface of the spheres [3, 19, 20].

The South African coal is sub-bituminous and generates coal fly ash that has low Fe content and the aqueous extracts of this coal fly ash are strongly alkaline (pH 12–12.5) due to the free lime content [16, 21–24]. The highly soluble CaO occurring as sub-micron fragments on the fly ash particles accounts for the alkaline properties of the coal fly ash [3].

Traditionally, AMD neutralization and remediation has been carried out using liming agents such as limestone, lime and sodium hydroxide [9, 10, 13, 25, 26]. An important disadvantage of these liming agents is cost and generation of large volumes of sludge that requires disposal. Due to those limitations, mining companies are always on the lookout for cheaper and effective liming agents.

Several research reports have highlighted the application of fly ash for the control of acid generation from sulphidic rich wastes [27, 28] and for amendment of acidic soils [29]. The studies involved blending the mine spoils, tailings and acidic soils with varying amounts of fly ash. These authors observed that the coal fly ash controlled the acid generation from the wastes by release of alkalinity over time to the system which neutralized the acidity produced. Several authors have reported on the successful application of coal fly ash for inorganic contaminants removal from AMD and acidic leachates [8, 19, 30, 31]. They observed that contaminants removal was directly proportional to CaO content of the fly ash. The authors concluded that coal fly ash due to its free lime content can be used as a neutralization and inorganic contaminants removal for AMD remediation.

Most of the power utilities generating coal fly ash in South Africa are located near coal mines supplying them with coal [6]. These coal mines are also the sources of AMD that requires remediation. This chapter reports on work done to explore the possibility of utilizing coal fly ash for treatment and remediation of AMD. The fact that most of the coal-powered utilities are near the coal mines that supply them with coal and produce AMD, makes the proposed treatment process economically viable. A fundamental understanding of the solution chemistry and product water quality after neutralization of AMD with coal fly ash is a prerequisite for this treatment process.

2. Collection of samples, physicochemical analysis and experimental methods

2.1. Collection and preparation of samples

Coal fly ash (CFA) samples were collected from two coal-powered utilities in Mpumalanga, South Africa and stored in tightly lockable plastic containers. The AMD samples used were collected from two collieries and a government AMD treatment plant that remediates AMD seepage from an old abandoned mine in Witbank, South Africa. The AMD samples from the government AMD treatment plant were scooped from the seepage point while samples for the two collieries consisted of AMD pumped from underground old mine workings to a collection dam and underground mine voids, respectively.

Raw AMD samples were filtered by using 0.45 μm pore cellulose nitrate membrane filters and diluted with MilliQ (ultrapure) water to EC < 1.5 mS/cm and 3 drops of dilute HNO_3 added and kept refrigerated at 4°C until analysis for cationic species. Samples for anion analysis were diluted with MilliQ water and refrigerated at 4°C until analysis.

2.2. Coal fly ash/acid mine drainage treatment experiments

The AMD/coal fly ash treatment experiments were designed to develop neutralization patterns that would indicate buffering properties of the AMD and also show the contact time required for the breakthrough to circum-neutral-alkaline pH in the product. Department of Water Affairs and Forestry [32] South Africa requires pH for treated water for release into surface water bodies to be in the range of 6–9. The batch treatment experiments were conducted by stirring a mixture of coal fly ash and AMD which was pre-determined to give a specific coal fly ash/AMD ratio (FA:AMD). An overhead stirring system was used to agitate the mixture for all experiments. The progress of the reaction was monitored by measuring the pH and EC with a Hanna HI 991301 portable pH/EC/TDS/Temperature metre.

A second set of treatment experiments were repeated at selected FA:AMD ratios to explore the product water chemistry and inorganic contaminants removal efficiency of the coal fly ash with increasing pH of the process water. The solid residues were separated by filtration and the process water samples were prepared for cationic and anionic species analysis. Cationic species were analyzed using ICP-MS (ELAN 6000) and the accuracy of the analysis monitored by the use of NIST water standards. Fe^{2+}/Fe^{3+} analysis was done by the colorimetric method using 2, 2-bipyridal as the complexing reagent. Sulfate analysis was done turbidimetrically by a portable data logging spectrophotometer (Hach DR/2010) and ion chromatography (Dionex DX-120).

Chemical characteristics of the coal fly ash samples were ascertained by X-ray fluorescence spectroscopy (XRF) by fusing with lithium metaborate. The solid residues collected after reacting coal fly ash with AMD were analysed by Phillips PANalytical X-ray diffractometer (XRD) using Cu Kα radiation generated at 20 mA and 40 KV. Specimens were step scanned as random powder mounts from 5 to 85° 2θ integrated at 0.02° 2θ per second. Powder samples of the solid residues were also observed under a scanning electron microscope (SEM-EDX) equipped with an energy dispersive X-ray analysis system (Hitachi X-650 microanalyzer).

3. Results and discussion

3.1. Chemical composition of coal fly ash

The chemical composition of the coal fly ashes used in the AMD treatment experiments are presented in **Table 1**.

The three major oxides identified in the two coal fly ashes were Al_2O_3, Fe_2O_3 and SiO_2. Coal fly ash B had higher Al_2O_3 content while coal fly ash A showed higher Fe_2O_3 content. These

Coal fly ash A				Coal fly ash B			
Species	**% (w/w)**	**Species**	**ppm**	**Species**	**% (w/w)**	**Species**	**ppm**
SiO_2	53.4 ± 2.4	Cu	47.3 ± 6.6	SiO_2	53.8 ± 0.29	Cu	57.9 ± 9.9
TiO_2	1.34 ± 0.05	Mo	5.22 ± 0.14	TiO_2	1.44 ± 0.11	Mo	6.56 ± 0.15
Al_2O_3	23.4 ± 1.1	Ni	93.4 ± 6.5	Al_2O_3	26.2 ± 2.52	Ni	58.2 ± 1.2
Fe_2O_3	4.72 ± 0.96	Pb	56.4 ± 13.6	Fe_2O_3	3.40 ± 0.24	Pb	29.1 ± 7.18
MnO	0.06 ± 0.002	Sr	1463.9 ± 111.8	MnO	0.05 ± 0.02	Sr	2056.0 ± 205
MgO	2.69 ± 0.05	Zn	57.3 ± 4.71	MgO	2.48 ± 0.58	Zn	25.4 ± 1.35
CaO	8.43 ± 0.57	Zr	488.1 ± 125.7	CaO	8.50 ± 1.75	Zr	536.1 ± 131.3
Na_2O	0.35 ± 0.25	Co	18.3 ± 13.08	Na_2O	0.49 ± 0.05	Co	10.4 ± 3.3
K_2O	0.49 ± 0.03	Cr	179.2 ± 1.14	K_2O	0.86 ± 0.07	Cr	122.7 ± 27.8
P_2O_5	0.35 ± 0.22	V	147.4 ± 38.9	P_2O_5	0.60 ± 0.22	V	145.8 ± 32.8
Cr_2O_3	0.03 ± 0.009	Ba	928.0 ± 91.9	Cr_2O_3	0.03 ± 0.006	Ba	1559.2 ± 346.7
NiO	0.011 ± 0.001			NiO	0.009 ± 0.002		
$V2O_5$	0.019 ± 0.002			V_2O_5	0.02 ± 0.003		
ZrO_2	0.052 ± 0.012			ZrO_2	0.055± 0.003		
LOI	2.36 ± 0.19			LOI	1.33 ± 0.36		

Table 1. Chemical composition of coal fly ashes A and B used in AMD treatment experiments.

coal fly ashes can be classified as class F according to the American Society for Testing and Materials [33] since ($SiO_2 + Al_2O_3 + Fe_2O_3 \geq 70\%$). The CaO content shows slight variation with coal fly ash B showing a slightly higher value. The total CaO content detected by XRF includes total CaO including that locked up in the aluminosilicate matrix. XRF does not distinguish the free lime from that trapped within the glass matrix. The free CaO content of CFA is important because of its availability for rapid dissolution, which has implications on its neutralization potential during AMD treatment. Substantial MgO was observed in both coal fly ashes and could contribute significantly in the neutralization of AMD. Both coal fly ashes showed high concentrations of Sr, Ba, Cr, Zr and Ni (**Table 1**). Trace elements of Mo are also present. These species are likely to be released into solution during the AMD treatment process.

3.2. Physicochemical properties of acid mine drainage samples

The physicochemical properties of the AMD samples used in the treatment experiments are presented in **Table 2**.

The AMD samples are highly acidic with the pH ranging from 2.39 for Navigation samples to 2.92 for Brugspruit samples (**Table 2**). All the samples exhibit high electrical conductivity

Parameter	Navigation	Bank	Brugspruit
pH	2.39 ± 0.05	2.46 ± 0.03	2.91 ± 0.02
EC (mS/cm)	10.83 ± 0.13	10.78 ± 0.15	10.02 ± 0.06
Acidity (mg/l $CaCO_3$)	6950 ± 70.7	7000 ± 70.7	500 ± 0.0
TDS (mg/L)	16765 ± 50.5	19410 ± 76.8	8975 ± 60.5
B	1.37 ± 0.163	1.51 ± 0.08	2.29 ± 0.221
Na	358.7 ± 2.95	399.9 ± 21.3	4137.9 ± 233.0
Mg	2661.7 ± 35.0	2844.2 ± 148.1	388.7 ± 19.7
Al	1068.1 ± 11.28	1140.1 ± 61.58	60.0 ± 2.9
Si	82.01 ± 1.24	87.8 ± 5.87	69.7 ± 3.5
K	23.03 ± 2.86	19.3 ± 4.21	52.6 ± 3.6
Ca	653.3 ± 10.6	1012.3 ± 75.9	842.1 ± 117.4
Mn	226.3 ± 4.7	242.3 ± 12.9	31.6 ± 1.50
Fe	5599.9 ± 80.9	6115.9 ± 327.5	250.8 ± 11.2
Fe^{2+}	3725.1 ± 30.5	2886.3 ± 20.7	153.1 ± 9.5
Fe^{3+}	1451.9 ± 45.2	3344.6 ± 50.5	126.1 ± 6.5
Ni	6.95 ± 0.02	7.96 ± 1.16	2.35 ± 0.13
Cu	0.355 ± 0.007	0.345 ± 0.018	0.116 ± 0.012
Co	4.3 ± 0.11	4.57 ± 0.32	1.15 ± 0.063
Zn	48.99 ± 30.63	17.7 ± 0.65	9.52 ± 0.49
Sr	7.69 ± 0.226	8.39 ± 0.45	1.05 ± 0.058
Mo	0.04 ± 0.002	0.044 ± 0.003	0.036 ± 0.004
Ba	0.209 ± 0.002	0.189 ± 0.01	0.148 ± 0.018
SO_4^{2-}	11888.1 ± 20.6	14949.7 ± 28.3	6155 ± 54.3
Cl^-	729.3 ± 15.3	265.9 ± 10.6	720 ± 11.5
NO_3^-	163.2 ± 21.6	41.6 ± 5.7	BDL

Table 2. Physicochemical properties of AMD samples used in the treatment experiments.

(10.02–11.36 mS/cm), typical of most AMD impacted mine waters. This is attributed to high sulfate content which contributes to the high conductivity values of the samples. The sulfate correlated positively with the EC measurements for most of the acidic coal mine waters investigated in this study. The sulfate recorded in these samples ranged from 6155 to 14950 mg/L making this anion dominant in the AMD samples while major cationic species included Na, Ca, Mg, Al, Mn and Fe. It is worth noting that Navigation and Bank AMDs had significantly higher total Fe iron content than Brugspruit AMD.

3.3. Treatment of Fe-rich acid mine drainage samples using different coal fly ashes: the role of coal fly ash chemistry

The pH and electrical conductivity (EC) trends of the treatment reactions at a FA to AMD ratio of 1:3 are shown in **Figure 1**.

The EC for treatment processes followed the same trend with continuous decrease as the agitation progressed. A zone was observed for the three treatment processes where the EC remained constant (**Figure 1**). In terms of contact time, it coincided with the pH buffer zones.

Both coal fly ashes A and B showed potential capacity to neutralize AMD at CFA:AMD ratios of 1:3. The pH-neutralization trends for all reactions showed a strong pH buffering zone at pH 6.2–6.8 which was observed to last for different reactions times depending on the coal fly ash or AMD type used. This pH buffer zone is associated with the oxidation and hydrolysis of Fe^{2+} which releases H^+ ions and delays the rise in pH [6] (Eq. (3)). Al^{3+} undergoes hydrolysis at pH ≈ 4.5 and will also contribute to pH buffer in this zone (Eq. (2)).

$$Al(H_2O)_6^{3+} \Rightarrow Al(OH)_3 + 3H^+ \quad \text{Acidity release : pH decrease} \tag{2}$$

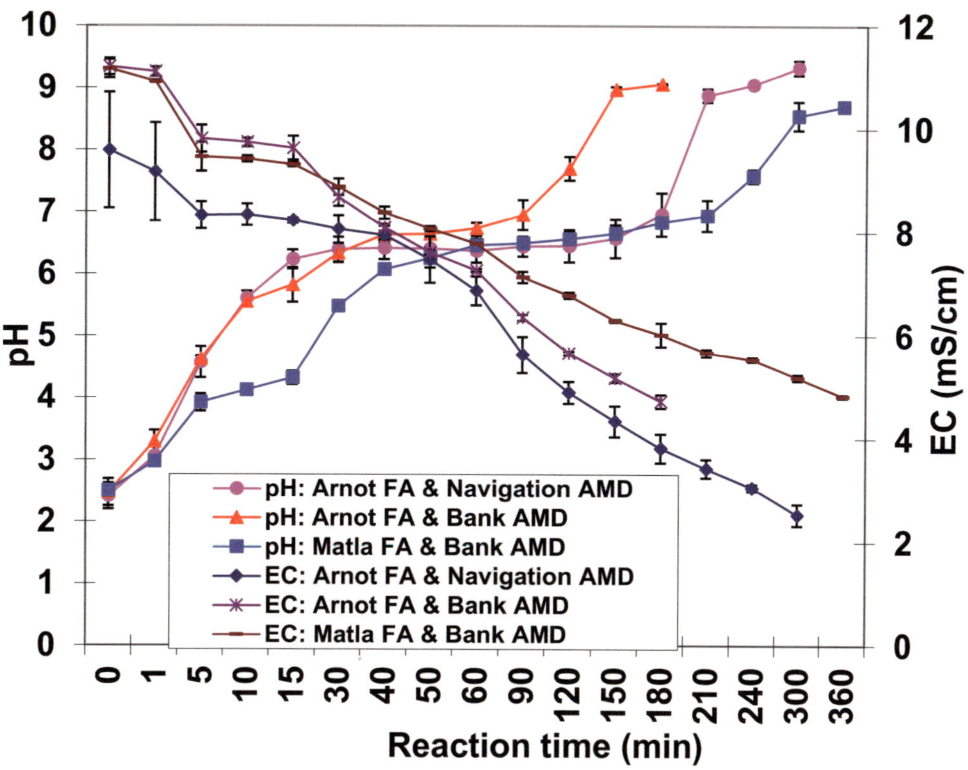

Figure 1. pH and EC for the treatment reactions at a FA:AMD ratio of 1:3 between, CFA-A and B with Bank AMD; and CFA-A with Navigation AMD. Values reported as mean ± SD (n = 4), error bars reflect 1 SD above and below the mean.

$$Fe^{2+} + 0.25O_2 + H^+ \Rightarrow Fe^{3+} + 0.5H_2O$$
$$Fe^{3+} + 3\,H_2O \Rightarrow Fe(OH)_3 + 3H^+ \text{ net acidity release: pH decrease} \tag{3}$$

This pH buffering zone is a result of high concentrations of Fe^{2+} and Al^{3+} in the AMD (≈3000–4000 mg/L for Fe^{2+} and ≈1150 mg/L for Al^{3+}) (**Table 2**). It should be noted in **Figure 1** that it took longer time for CFA-A to overstep the acidic buffer zone with Navigation AMD as compared to Bank AMD. This is attributed to the high concentrations of Fe^{2+} in Navigation AMD than Bank AMD (**Table 3**) that could sustain the buffering capacity. It is observed that both solutions broke through to a pH of ≈9, with Navigation AMD taking 200 min and only 150 min for Bank AMD to break through. This indicates that the chemistry of the AMD is crucial in this treatment process. A small buffer zone was also observed at pH 4–4.5 for treatment reaction between CFA-B and Bank AMD. This buffer zone is attributed to the hydrolysis and precipitation of Fe^{3+} with subsequent consumption of H^+ (Eq. (3)). It should be observed that Bank AMD had twice the concentration of Fe^{3+} as Navigation AMD. This aspect is also confirmed in the pH trend for the CFA-A and Bank AMD with a slight buffering at pH ≈5.5 (**Figure 1**). Coal fly ash B appeared to have less free alkalinity than CFA-A. This is evident from the pH trends during the treatment process between both CFAs and Bank AMD. To break through to pH ≥ 8.7 (**Figure 1**), contact time required was greater with CFA-B than CFA-A for the treatment of Bank AMD. The neutralization potential of CFA depends on free CaO available and in this case, CFA-A seems to have more free CaO than CFA-B. It should be noted that XRF gives the total CaO content of the coal fly ash and therefore results in **Table 1** which do not reflect free CaO. This aspect confirms that the chemistry of the coal fly ash will also be an important factor in this treatment process.

3.4. Treatment of iron-rich and iron-poor acid mine drainage with coal fly ash: the role of CFA/AMD ratios and acid mine drainage chemistry

To evaluate the effect of AMD chemistry and CFA/AMD ratios on the treatment process and product water quality, CFA-B was reacted with Brugspruit and Navigation AMD. Brugspruit AMD had low total Fe, Al and Mn content while Navigation AMD had high concentration of total Fe, Al and Mn in addition to high SO_4^{2-} content (**Table 2**). **Figure 2** shows the trends of pH and electrical conductivity (EC) for the treatment process between CFA-B and Brugspruit AMD at various CFA:AMD ratios.

The treatment of Brugspruit AMD with low ratios of CFA-B led to alkaline pH (pH ≈ 10) within 5 min of reaction time (**Figure 2**). This indicates that even low amounts of FA could be used to achieve neutralization in some cases depending on the chemistry of AMD being treated. Treatment of Brugspruit AMD with CFA-B:AMD ratios of between 1:3.5 and 1:8 led to highly alkaline pH (pH > 12) in product water. The ratio of CFA used was directly proportional to the decrease of EC. The EC seemed to decrease within the initial 5 min of treatment and then stabilized at a minimum of 8.5–9 mS/cm. It should be observed all the CFA-B:AMD treatment ratios used led to a final pH ≥ 12 after 360 min of reaction. It should be noted that the buffer zones observed with CFA-A:Navigation AMD and CFA-B:Bank AMD treatment are not observed with Brugspruit AMD in all the CFA:AMD ratios employed (**Figures 1** and **2**). This

http://dx.doi.org/10.5772/intechopen.69741

													DWAF limits	
FA:AMD	AMD		1:3.5		1:5		1:8		1:2		1:3		Irrigation	Domestic use
Species	ave	SD	ave	SD	ave	SD	ave	SD	ave	SD	ave	SD		
B	2.29	0.221	1.92	0.123	2.79	0.004	5.93	0.57	6.12	0.59	4.42	1.35	0.5	0.05
Na	4137.99	233.007	1482.34	50.64	1561.34	55.04	1580.24	45.5	3034.67	80.84	2394.18	100.4	70	100
Mg	388.65	19.707	0.24	0.021	1.46	0.21	0.833	0.021	297.04	21.4	236.86	25.3		30
Al	60.04	2.886	0.15	0.003	0.99	0.06	2.44	0.25	0.153	0.011	0.32	0.015		
Si	69.73	3.495	0.82	0.013	2.65	0.28	3.29	0.201	13.99	2.91	16.72	1.25		
K	52.59	3.634	20.21	3.74	21.26	1.34	15.54	2.5	45.33	5.69	37.53	5.43		50
Ca	842.11	117.44	546.89	60.34	478.65	20.56	793.38	30.58	877.22	20.56	635.68	30.98		32
Cr	0.77	0.067	0.5	0.034	0.526	0.023	0.375	0.028	0.525	0.021	0.176	0.022		
Mn	31.58	1.481	0.028	0.002	0.066	0.013	0.026	0.002	0.339	0.004	6.24	0.68		
Fe	250.84	11.203	2.62	0.421	3.497	0.251	0.853	0.037	4.23	0.15	3.52	0.15		
Ni	2.35	0.127	0.031	0.002	0.051	0.002	0.194	0.022	0.332	0.002	0.602	0.021	0.2	0–0.02
Co	1.15	0.063	0.001	0.0005	0.002	0.0005	0.004	0.00057	0.017	0.001	0.115	0.002	0.05	0–0.05
Cu	0.11	0.011	0.031	0.001	0.04	0.002	0.049	0.003	0.044	0.002	0.045	0.0015	0.2	1
Zn	9.52	0.491	1.13	0.052	9.19	1.19	16.44	1.48	1.21	0.08	8.047	0.038	1	3
As	0.11	0.491	0.004	0.001	0.006	0.001	HDL		0.014	0.002	0.007	0.002		
Se	0.032	0.006	0.072	0.003	0.054	0.003	0.084	0.018	0.04	0.018	0.017	0.001		
Sr	1.046	0.057	13.86	2.54	9.539	2.31	9.23	2.68	10.23	1.58	5.71	0.32		
Mo	0.036	0.0037	0.219	0.002	0.181	0.011	0.16	0.004	0.134	0.013	0.092	0.003	0.01	0–0.05
Cd	0.012	0.0007	0.002	0.0004	0.002	0.001	0.003	0.00035	0.001	0.0005	0.002	0.0005		
Ba	0.148	0.0147	0.569	0.005	0.684	0.026	0.626	0.058	0.579	0.026	0.432	0.015		

													DWAF limits	
FA:AMD	AMD		1:3.5		1:5		1:8		1:2		1:3		Irrigation	Domestic use
Pb	0.178	0.02	0.015	0.002	0.017	0.002	0.018	0.003	0.011	0.005	0.013	0.001		
SO_4	6165	10	6137	7	5668.03	47.53	4601.77	3.37	7182.96	7.46	3709.80	58.99		
Cl	720	22.3	385.28	19.5	285.52	6.8	276.15	20.3	449.73	15.8	326.12	19.2		
pH	2.55	0.12	9.16	0.04	9.73	0.13	12.04	0.05	12.64	0.14	12.62	0.12		6–9

ave, average; SD, standard deviation.

Table 3. Concentration of inorganic contaminants at various final pH of solution and CFA:AMD ratios for treatment of Brugspruit AMD with CFA-B (concentrations are in mg/L except for pH).

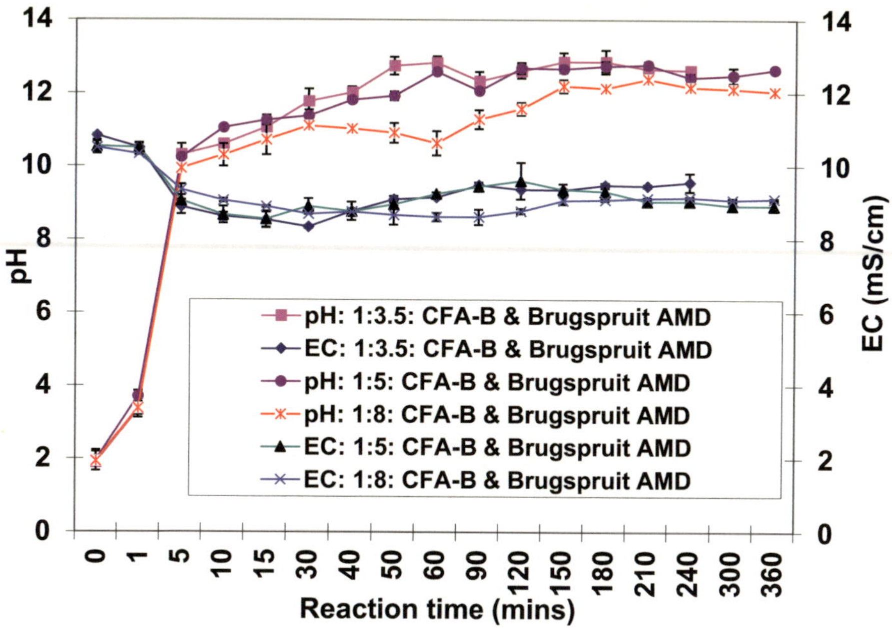

Figure 2. pH and EC trends for the treatment process for various CFA-B:Brugspruit AMD ratios. Values reported as mean ± SD ($n = 4$), error bars reflect 1 sample SD above and below the mean.

is a strong indication that AMD chemistry has a significant role to play in the final product water chemistry.

Figure 3 shows the trends of pH and electrical conductivity (EC) for the treatment process between CFA-B and Navigation AMD at various CFA:AMD ratios.

The pH and EC trends for the treatment process of Navigation AMD with CFA-B are distinctively different from those of the treatment process for Brugspruit AMD (**Figures 2** and **3**). For the Navigation AMD treatment process, there is a stepwise increase in pH as the treatment progresses. Buffer zones are also observed at pH 3.5–4 and pH 5.5–6.5.

Despite the high CFA:AMD ratios employed in the treatment as compared to Brugspruit treatment, the breakthrough to pH > 10 only occurs after 210 min (**Figure 3**) and for lower CFA:AMD ratio of 1:3, a breakthrough was not achieved after 360 min. It should be noted that the stepwise increase in pH with time is lacking for the Brugspruit AMD. Stepwise and gradual decrease in EC is noted for the Navigation AMD treatment process (**Figure 3**). The initial decrease in EC for the Brugspruit treatment is not sustained and stabilizes at 8.67–9.56 mS/cm. Uhlmann et al. [34] observed that the buffering observed at pH 3.5–4 is due to hydrolysis of Fe^{3+} while oxidation and hydrolysis of Fe^{2+} contributes greatly to buffering at pH 5.5–6.5 [25] (Eqs. (2) and (3)). At pH ≥ 4.5, Al^{3+} undergoes hydrolysis forming insoluble hydroxides and will buffer pH in the region of 3.5–4 (Eq. (2)) [34, 35]. It should be noted that Brugspruit AMD had very low concentration of Fe^{3+}/Fe^{2+} and Al^{3+} as compared to Navigation AMD (**Table 2**). This could explain the lack of stepwise decrease in pH and buffering during the treatment process. There are two

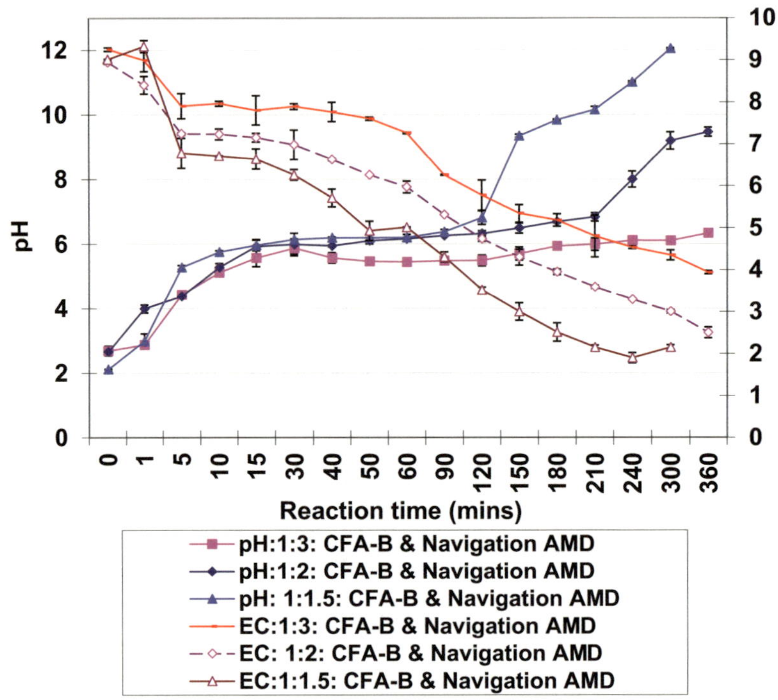

Figure 3. pH and EC for various FA:AMD ratios for reaction between CFA-B and Navigation AMD. Values reported as mean ± SD (*n* = 4), error bars reflect 1 SD above and below the mean.

opposing processes that finally control the final pH of the product water in this treatment process. The dissolution and hydrolysis of soluble alkaline oxides such as CaO and MgO (**Table 1**) from coal fly ash will contribute to increase in pH (Eqs. (4) and (5)). This is also confirmed by the fact that the time taken for breakthrough of the buffer zone 5.5–6.5 reduced with increase of CFA in the treatment mixture. CFA:AMD ratio of 1:1.5 took the shortest time to breakthrough this buffer zone (**Figure 3**).

$$CaO + H_2O \Rightarrow Ca^{2+} + 2OH^- \text{ pH increase} \tag{4}$$

$$MgO + H_2O \Rightarrow Mg^{2+} + 2OH^- \text{ pH increase} \tag{5}$$

Hydrolysis of AMD constituents such as Fe^{3+}, Al^{3+} and Fe^{2+} releases protons (acidity increase) (Eqs. (2) and (3)) and offsets the pH increase attributed to the dissolution of the alkaline oxide from coal fly ash. The net result is a pH buffer zone, this buffer zone is only overstepped when the components leading to release of acidity are completely hydrolyzed. These hydrolysable components will finally determine if the final pH of the product water will be acidic or alkaline at any given treatment time. It will also be observed that the final pH of the product water also depended on the CFA:AMD ratio applied. Higher CFA:AMD led to higher pH of product water regardless of treatment time. It can therefore be concluded that factors that will control the product water chemistry and quality would be, CFA:AMD ratio, the treatment time and the chemistry of the AMD.

3.5. Inorganic contaminants removal: product water quality compared to South African water quality guidelines

Acid mine waters are highly reactive solutions that can dissolve most primary minerals when reacted with an alkaline solid waste material such as coal fly ash with subsequent formation of a variety of secondary minerals that can potentially adsorb trace chemical species. Dissolution of coal fly ash by AMD triggers several processes such as hydrolysis of soluble alkaline oxides leading to increase in pH. Several authors observe that pH is the most important parameter in coal fly ash solutions and accounts for significant inorganic toxic elements removal through precipitation, ion exchange and adsorption processes [13, 30].

Tables 3 and **4** present results of inorganic contaminants removal as a function of final solution pH and CFA:AMD ratios for Brugspruit and Navigation AMD treatment with CFA-B. The results are discussed with respect to pH of precipitation of the various metal species as determined from thermodynamic calculations and experimental observations from titration of solutions containing the stated species as reported by Britton [36].

3.6. Brugspruit AMD treatment with CFA-B: inorganic contaminants removal

Brugspruit AMD had different chemistry compared to that of Navigation AMD, it had low concentration of the major hydrolysable chemical species Al^{3+}, Fe^{3+}, Fe^{2+} and Mn^{2+}. Major inorganic contaminants Fe, Mn, Al and Mg were reduced significantly in all the ratios investigated (**Table 3**). Fe and Al were reduced by >95% at pH ≥ 9.16 attained in the final process waters for CFA:AMD ratios, 1:3, 1:5, 1:8, 1:2 and 1:3. It should be noted that final pH also depended on the treatment time employed. Mn removal was ≥90% for all CFA:AMD ratios evaluated except for 1:3 ratio at 80%. Jenke et al. [25] observed that at the pH of minimum solubility of the hydroxides of Fe^{3+} (pH 3.0), of Fe^{2+} (pH 6.0–8.0), of Mn^{2+} (pH 8.41–9.0) and of Zn^{2+} (pH 6.0–6.5), a significant proportion of the initial concentration should be precipitated out of solution. Gitari et al. [19] observed that at pH ≥ 4.5, most of the mineral phases bearing these species were at saturation or oversaturation (SI ≥ 1). At pH ≥ 12, a slight increase in Al, Fe and Mn was observed. At pH 12.0–12.5 attained for ratios 1:8, 1:2 and 1:3, the formation of hydroxy complexes was attributed to the decreased removal. Drever [37] observes that Al^{3+} hydroxide exhibits minimum solubility at pH 6–6.5, therefore at pH > 6.5, the Al-hydroxy complexes become more important [38].

Among the minor and trace species, Ni, Cu, Pb, Mg, Cr, Ni,Co, Cu, As and Cd were all observed to decrease as the pH increased. The Cu, Cr, Zn and Pb removal efficiency was observed to decrease at CFA:AMD ratio of 1:8 which attained a pH of 12.04. The decrease in removal of these species could be attributed to soluble hydroxo species formed at pH > 12.0. Another reason for the increase could be reduced adsorption due to reduced formation of amorphous $Fe(OH)_3$, MnOOH and $Al(OH)_3$ due to the low concentration of Fe^{3+}, Mn and Al^{3+} in Brugspruit AMD. Mg^{2+} removal efficiency approached 100% at pH 9.16 with a decrease observed as the pH increased to 12.62. Decreased removal could be due to the formation of

FA:AMD	AMD		1:3		1:2.5		1:2		1:1.5		DWAF limits	
Species	ave	SD	ave	SD	ave	SD	ave	SD	ave	SD	Irrigation	Domestic use
B	1.37	0.163	23.44	3.44	24.38	2.56	17.57	2.356	17.86	3.456	0.5	0–0.5
Na	358.72	2.946	71.76	10.23	72.53	9.87	68.21	10.12	62.02	21.8	70	100
Mg	2661.67	35.008	636.85	80.45	618.06	70.8	200.04	34.5	1.5	0.02		30
Al	1068.00	11.279	2.85	0.05	3.26	0.35	2.35	0.56	9.41	1.97		
Si	82.01	1.238	4.39	0.52	2.05	0.003	2.3	0.16	1.09	0.03		
K	23.03	2.856	16.27	2.45	19.45	1.56	14.87	2.35	10.33	2.45		50
Ca	653.33	10.626	368.19	23.8	495.11	20.45	448.53	19.67	477.92	35.63		32
Cr	1.11	0.009	0.069	0.003	0.163	0.013	0.141	0.001	0.083	0.015		
Mn	226.2.5	4.742	56.65	5.68	5.15	0.456	1.11	0.005	0.133	0.005		
Fe	5599.92	80.862	293.3	19.57	52.25	6.78	43.23	4.578	4.7	0.67		
Ni	6.95	0.018	0.58	0.04	0.134	0.002	0.088	0.004	0.051	0.001	0.2	0–0.02
Co	4.3	0.105	0.312	0.012	0.014	0.001	0.019	0.003	0.002	0.0015	0.05	0–0.05
Cu	0.355	0.007	0.045	0.003	0.055	0.011	0.073	0.005	0.034	0.006	0.2	1
Zn	48.99	30.624	1.3	0.012	1.2	0.05	1.26	0.02	0.736	0.012	1	3
As	0.193	0.012	0.003	0.0005	0.005	0.0015	0.003	0.0005	0.004	0.001		
Se	0.032	0.002	0.037	0.002	0.026	0.003	0.052	0.001	0.112	0.003		
Sr	7.69	0.226	15.71	1.34	17.18	2.543	14.48	2.87	17.72	1.48		
Mo	0.04	0.002	0.025	0.002	0.622	0.002	0.665	0.002	0.77	0.04	0.01	0–0.05
Cd	0.032	0.001	0.003	0.0015	0.004	0.001	0.003	0.0015	0.002	0.0004		
Ba	0.209	0.002	0.369	0.003	0.347	0.032	0.336	0.034	0.319	0.013		
Pb	0.314	0.107	0.019	0.002	0.041	0.004	0.0301	0.005	0.0154	0.0021		

FA:AMD	AMD		1:3		1:2.5		1:2		1:1.5		DWAF limits	
Species	**ave**	**SD**	**ave**	**SD**	**ave**	**SD**	**ave**	**SD**	**ave**	**SD**	**Irrigation**	**Domestic use**
SO_4	11949.6	61.5	5483.3	14.9	2414.3	28.1	2508.1	247.6	4570.7	110.3		
NO_3	163.17	10.01	68.43	7.82	23.69	5.34	<0.1	0	9.3.44	10.21		
Cl	729.27	100.1	65.8	3.21	63.168	6.21	44.744	3.03	88.172	9.65		
pH	2.69	0.21	6.33	0.15	8.72	0.62	9.47	0.43	12.1	0.18		6–9

ave, average; SD, standard deviation.

Table 4. Concentration of inorganic contaminants at various final pH of solution and CFA:AMD ratios for treatment of Navigation AMD with CFA-B (concentrations are in mg/L except for pH).

soluble hydroxo species at the high pH. Na and K were removed by 50% as the pH increased. Gitari et al. [19] observed that as pH increases and in presence of SO_4^{2-}, several Na-K bearing mineral phases such as jarosite are precipitated and would account for this decrease. Sulfate did not register significant removal and in some CFA:AMD ratios (1:2) was observed to increase, this increase is attributed to release from the coal fly ash.

3.7. Navigation AMD treatment with CFA-B: inorganic contaminants removal

The pH increase with increase of CFA:AMD ratios was observed to be gradual for Navigation AMD treatment as compared to that of Brugspruit AMD (**Tables 3** and **4**). Several inorganic contaminants removal trends are observed as the pH increases. At final pH 6.33 achieved for the CFA:AMD ratio of 1:3, a sharp decrease in concentration is observed for Fe, Al and Mn. These are the major inorganic contaminants that characterize Navigation AMD. With subsequent CFA:AMD ratios, the removal efficiency of ≥90% are observed. Gitari et al. [19] observed that as pH increases, mineral phases bearing these chemical species are precipitated and account for their increased removal in solution. Minor and trace inorganic species such as Si, Cr, Ni, Co, Cu, Zn, As, Se, Cd and Pb were all observed to decrease as the pH increased with subsequent CFA:AMD ratios. Their removal could be attributed to several mechanisms such as precipitation of mineral phases and adsorption to high surface area Fe and Al oxyhydroxy precipitates. Gitari et al. [19] observed that co-precipitation of Fe, Al-oxyhydroxides, oxyhydroxy sulfates and adsorption could be responsible for attenuation of Cu^{2+}, Zn^{2+} species.

Britton [36] observes that the pH of minimum solubility of the hydroxides/oxyhydroxides of Fe^{3+}, Fe^{2+}, Al^{3+}, Zn^{2+}, Cu^{2+} and Ni^{2+} are 3.0, 6.0–8.0, 4–4.5, 6.0–6.5 and 6.66, respectively. The pH attained for CFA:AMD ratio of 1:3–1:2 is within this range. This would explain the high removal of these species within this pH range. However, Cu and Pb registered an increase in concentration in final process waters for CFA:AMD 1:2 and 1:2.5 ratios whose final pH ranged from 8.7 to 9.5 (**Table 4**). This would be attributed to the formation of soluble hydroxo complexes for both species at this pH range.

Mg, Na, K and Ca removal increased with increasing pH of the final process water. These chemical species are highly soluble in the pH range generated in process waters and likely removed through incorporation in precipitating mineral phases. Gitari et al. [19] observed that mineral phases such as alunite, basaluminite, jurbanite, brucite and jarosite are likely to precipitate out in CFA:AMD reactions leading to removal of these species. Sulfate was likely removed through formation of gypsum, incorporation into Al, Fe-oxyhydroxysulfates. A direct relationship between percent sulfate removal and FA content in the reaction mixture suggests that dissolution of CaO and subsequent formation of gypsum accounts for the sulfate removal in both AMD treatments. XRD analysis identified gypsum in the solid residues of the CFA:AMD reactions [3, 39].

3.8. Leaching of chemical species from coal fly ash in the treatment process

Coal fly ash is a reactive material and on contact with aqueous solutions such as AMD undergoes dissolution with subsequent release of chemical species. An observation of results in **Tables 3** and **4** show that interaction of coal fly ash and AMD leads to the release of B, Sr, Mo

http://dx.doi.org/10.5772/intechopen.69741

and Ba. The fact that these chemical species increased with increase in the ratio of CFA in the treatment mixture attests to the fact they were being released from the coal fly ash. Mattigod et al. [17] and Eary et al. [20] reported that the glassy phase in coal fly ash is enriched with As, Na, Mg, K, Sr, B, Mo and Ba as soluble salts which are released into solution when the CFA contacts aqueous solutions.

3.9. Comparison to South Africa water quality guidelines

The product water for various CFA:AMD ratios was compared to the Department of Water Affairs and Forestry [32] water quality limits for irrigation and domestic use. Treatments of Navigation AMD with CFA-B resulted in much cleaner water with Cu, K, Mo, Na and Zn being within the domestic water use limits for 1:3 CFA:AMD ratio and Co, Cu, K, Na and Zn being within the domestic water limits for 1:2.5 CFA:AMD ratio. Treatment of Brugspruit AMD with CFA-B produced less clean water although breakthrough to alkaline pH was established within less than an hour. Only Cu and K were within the domestic limits for 1:3 FA:AMD ratio. This comparison includes only the FA:AMD ratios that resulted in process water in the pH range 6–9. The alkali and alkaline earth metal species (Na, B, Mg, Ca, Sr and Ba) remained largely in the process water in concentrations beyond the guideline limits. B, Mg, Sr, Mo and Ba were largely released from dissolution of coal fly ash and additional treatment options such as reverse osmosis will be required to further clean the water to required standards.

4. Conclusions

The dissolution and hydrolysis of basic oxides such as MgO and CaO from coal fly ash on contacting AMD led to an increase in pH. On the other hand, hydrolysis of AMD chemical species such as Al^{3+}, Fe^{3+}, Fe^{2+} and Mn^{2+} led to release of acidity that counteracted the pH increase leading to stepwise increase in pH of process water and buffer zones as the treatment processes progressed. This was clearly observed for Navigation AMD as opposed to Brugspruit AMD that had low hydrolysable species. These two processes initiated several processes such as precipitation and adsorption that led to attenuation of the major and minor chemical species leading to much cleaner product water. Efficiency of the inorganic contaminants removal was directly linked to amount of FA in the reaction mixture and to the final pH in product water. Three processes were identified that would finally control the quality of the product water, these were (1) CFA:AMD ratios, (2) treatment time and (3) chemistry of the AMD. Most of the major contaminants were reduced to below South African water quality guidelines; however, the coal fly ash was observed to release other contaminants in solution and this could be a drawback for the proposed treatment process. Other factors that would improve the sustainability of the treatment process are beneficial application of the treatment residues as backfill material. Economic viability of this process would also depend on closeness of the AMD generating mines and the coal power utilities generating the coal fly ash.

Acknowledgements

The authors would like to acknowledge and thank Water Research Commission (WRC), National Research Foundation (NRF) and Coaltech 2020 for funding this research work.

Author details

Wilson Mugera Gitari[1]*, Leslie F. Petrik[2] and Segun A. Akinyemi[3]

*Address all correspondence to: mugera.gitari@univen.ac.za

1 Environmental Remediation and Water Pollution Chemistry Group, School of Environmental Sciences, University of Venda, Thohoyandou, South Africa

2 Environmental and Nano Science Research Group, Department of Chemistry, University of the Western Cape, South Africa

3 Department of Geology, Ekiti State University, Ado Ekiti, Nigeria

References

[1] Simate GS, Ndlovu S. Acid mine drainage: Challenges and opportunities. Journal of Environmental Chemical Engineering. 2014;**2**:1785-1803. http://dx.doi.org/10.1016/j.jece.2014.07.021

[2] Younger PL, Banwart SA, Hedin RS. Mine water chemistry. In: Brian JA, Jack TT, editors. Mine Water: Hydrology, Pollution, Remediation. Dordrecht: Kluwer Academic Publishers; 2002. pp. 65-117

[3] Gitari MW, Petrik LF, Etchebers O, Key DL, Iwuoha E, Okujeni C. Treatment of acid mine drainage with fly ash: Removal of major contaminants and trace elements. Journal of Environmental Science Health-Part A. 2006;**A41**:1729-1747. DOI: 10.1080/10934520600754425

[4] Feng D, Aldrich C, Tau H. Treatment of acid mine water by use of heavy metal precipitation and ion exchange. Mineral Engineering. 2000;**13**:623-642. http://dx.doi.org/10.1016/S0892-6875(00)00045-5

[5] Gunther P, Mey W, Van Niekerk A. A sustainable mine water treatment initiative to provide potable water for South African City _ a public-private partnership. Water in Mining Conference; 2006; Brisbane

[6] Gitari WM. Attenuation of metal species in acidic solutions using bentonite clay: Implications for acid mine drainage remediation. Toxicological & Environmental Chemistry. 2014;**96**:201-217. DOI: 10.1080/02772248.2014.923426

[7] Nordstrom DK. Aqueous pyrite oxidation and the consequent formation of secondary iron minerals. In Acid Sulfate Weathering (eds. J. A. Kittrick, D. F. Fanning, and L. R. Hossner). Soil Science Society of America Special Publication. 1982;**10**:37-56. ISBN-13: 978-0-08-044643-1

[8] Potgieter-Vermaak SS, PotgieterJH, Monama P, Van Grieken R. Comparison of limestone, dolomite and fly ash as Pre-treatment agents for acid mine drainage. Minerals Engineering. 2006;**19**:454-462. http://dx.doi.org/doi:10.1016/J.MINENG.2005.07.009

[9] Maree JP, du Plessis P, van der Watt CJ. Treatment of acidic effluents with limestone instead of lime. Water Science Technology. 1992;**26**:345-355

[10] Maree JP, Van Tonder GJ, Millard P. Underground Neutralization of Mine Water with Limestone. Pretoria, South Africa: Water Research Commission, Report no. 609/1/96; 1996. ISBN: 1-86845-241-7

[11] Allison AE, Nathoo J, Thomson K, Kramer HJ, Witkamp GJ, Reddy ST, Randall DG. Design of a eutectic freeze crystallization process for multicomponent waste water stream. Chemical Engineering Research and Design. 2010;**88**:1290-1296. http://dx.doi.org/10.1016/j.cherd.2010.01.023

[12] Mulopo J. Continuous pilot scale assessment of the alkaline barium calcium desalination process for acid mine drainage treatment. Journal of Environmental Chemical Engineering. 2015;**3**:1295-1302. http://dx.doi.org/10.1016/j.jece.2014.12.001

[13] Cravotta III AC, Trahan MK. Limestone drains to increase pH and remove dissolved metals from acidic mine drainage. Applied Geochemistry. 1999;**14**:581-606. http://dx.doi.org/10.1016/S0883-2927(98)00066-3

[14] Cogho VE,van Niekerk AM. Optimum coal mine water reclamation project. In: Proceedings of the International Mine Water Conference; 2009; Pretoria, South Africa. ISBN Number: 978-0-9802623-5-3

[15] Bhattacharyya KG, Gupta SS. Kaolinite and montmorillonite as adsorbents for Fe^{3+}, Co^{2+} and Ni^{2+} in aqueous medium. Applied Clay Science. 2008;**41**:1-9. http://dx.doi.org/10.1016/j.clay.2007.09.005

[16] Sammy MN, Chuks PE, Richard OA, Gitari WM, Segun AA, Olanrewaju OF, Petrik LF. The leaching behavior and geochemical fractionation of trace elements in hydraulically disposed weathered coal fly ash. Journal of Environmental Science and Health, Part A: Toxic/Hazardous Substances and Environmental Engineering. 2014;**49**:233-242. DOI: 10.1080/10934529.2013.838929

[17] Mattigod SV, Rai D, Eary LE, Ainsworth CC. Geochemical factors controlling the mobilization of inorganic constituents from fossil fuel residues: Review of the major elements. Journal of Environmental Quality. 1990;**19**:188-201. DOI: 10.2134/jeq1990.00472425001900020004x

[18] Gitari WM, Fatoba OO, Petrik LF, Vadapalli VR. Leaching characteristics of selected South African fly ashes: Effect of pH on the release of major and trace species.

Journal of Environmental Science and Health, Part A. 2009;**44**:206-220. DOI: 10.1080/10934520802539897

[19] Gitari WM, Petrik LF, Etchebers, Key DL, Okujeni C.Utilization of fly ash for treatment of coal mines wastewater: Solubility controls on major inorganic contaminants. Fuel. 2008;**87**:2450-2462. DOI: 10.1016/j.fuel.2008.03.018

[20] Eary LE, Dhanpat R, Mattigod SV, Ainsworth CC. Geochemical factors controlling the mobilization of inorganic constituents from fossil fuel combustion residues: II. Review of the minor elements. Journal of Environmental Quality. 1990;**19**:202-214. DOI: 10.2134/jeq1990.00472425001900020005x

[21] Foner HA, Robl TL, Hower JC, Graham UM. Characterization of fly ash from Israel with reference to its possible utilization. Fuel 1999;**78**:215-223. http://dx.doi.org/10.1016/S0016-2361(98)00140-9

[22] Eze CP, Sammy MN, Richard OA, Gitari WM, Segun AA, Olanrewaju OF, Petrik LF. Chemical, mineralogical and morphological changes in weathered coal fly ash: A case study of a brine impacted wet ash dump. Journal of Environmental Management. 2013;**129**:479-492. http://dx.doi.org/10.1016/j.jenvman.2013.07.024

[23] Grace NM, Leslie FP, Olanrewaju OF, Gitari WM, Frédéric JD, Jaco N, Sammy MN, Paul EC. Comparison of CO_2 capture by ex-situ accelerated carbonation and in in-situ naturally weathered coal fly ash. Journal of Environmental Management. 2013;**127**:212-220. http://dx.doi.org/10.1016/j.jenvman.2013.05.027

[24] Fatoba OO, Petrik LF, Akinyeye RO, Gitari WM, Iwuoha EI. Long-term brine impacted fly ash. Part 1: Chemical and mineralogical composition of the ash residues. International Journal of Environmental Science and Technology. 2015;**12**:551-562. DOI 10.1007/s13762-013-0439-1

[25] Jenke RD, Gordon KP. Chemical changes in concentrated, acidic, metal-bearing waste waters when treated with lime. Environmental Science and Technology. 1983;**17**:217-223. DOI: 10.1021/es00110a007

[26] Doye I, Duchesne J. Neutralization of acid mine drainage with alkaline industrial residues: Laboratory investigation using batch-leaching tests. Applied Geochemistry. 2003;**18**:1197-1213. http://dx.doi.org/10.1016/S0883-2927(02)00246-9

[27] Xenidis A, Evangelia M, Ioannis P. Potential use of lignite fly ash for the control of acid generation from sulphidic wastes. Waste Management. 2002;**22**:631-641. http://dx.doi.org/10.1016/S0956-053X(01)00053-8

[28] Seoane S, Leiros CM. Acidification-Neutralization processes in a lignite Mine Spoil Amended with Fly Ash or Limestone. Journal of Environmental Quality. 2001;**30**:1420-1431. DOI: 10.2134/jeq2001.3041420x

[29] Phung HT, Lund LJ, Page AL, Bradford GR. Trace elements in fly ash and their release in water and treated soils. Journal of Environmental Quality. 1979;**8**:171-175. DOI: 10.2134/jeq1979.00472425000800020007x

[30] Erol M, Kucukbayrak S, Ersoy-Mericboyu A, Ulubas T. Removal of Cu^{2+} and Pb^{2+} in aqueous solutions by fly ash. Energy Conservation Management. 2005;**46**:1319-1331. http://dx.doi.org/10.1016/j.enconman.2004.06.033

[31] Polat M, Eli L, Ithamar P, Haim C. Chemical neutralization of acidic wastes using fly ash in Israel. Journal of Chemical Technology and Biotechnology. 2002;**77**:377-381. DOI: 10.1002/jctb.567

[32] Department of Water Affairs and Forestry. South African Water Quality, Guidelines, second ed, Domestic Use. Vol. 1. 2nd ed. Pretoria, South Africa: Department of Water Affairs and Forestry; 1996. ISBN 0-7988-5339-5

[33] American Society for Testing and Materials. Standard Specification for Fly Ash and Raw or Calcined Natural Pozzolan for Use as a Mineral Admixture in Portland Cement Concrete. Philadelphia, PA: ASTM; 1988. pp. C618-88

[34] Uhlmann WH, Buttcher O, Totsche CE, Steinberg W. Buffering of acidic mine lakes: The relevance of surface exchange and solid bound sulphate. Mine Water and Environment. 2004;**23**:20-27. DOI: 10.1007/s10230-004-0032-4

[35] Fillipek HL, Nordstrom DK, Ficklin HW. Interaction of acid mine drainage with waters and sediments of west squaw creek in the West Shasta Mining District, California. Environmental Science and Technology, 1987;**21**:388-396. DOI: 10.1021/es00158a009

[36] Britton HTS. Hydrogen Ions: Their Determination and Importance in Pure and Industrial Chemistry. 4th ed. Vol. 2. London, UK:Chapman and Hall Ltd; 1956

[37] Drever IJ. The Geochemistry of Natural Waters: Surface and Groundwater Environments. 3rd ed. Prentice-Hall, Inc, New Jersey, United states; 1997. pp. 25-37. ISBN-13: 978-0132727907

[38] Shum M, Lavkulich L. Speciation and solubility relationships of Al, Cu and Fe in solutions associated with sulfuric acid leached mine waste rock. Environmental Geology. 1999;**38**:59-68. DOI: 10.1007/s002540050401

[39] Godfrey Madzivire, Leslie F. Petrik, Wilson M. Gitari, Tunde V. Ojumu, Gillian Balfour. Application of coal fly ash to circumneutral mine waters for the removal of sulphates as gypsum and ettringite. Minerals Engineering; 2010;**23**:252-257. http://dx.doi.org/10.1016/j.mineng.2009.12.004

Chapter 6

Phytoreclamation of Abandoned Acid Mine Drainage Site After Treatment with Fly Ash

Madhumita Roy, Roopali Roychowdhury,
Pritam Mukherjee, Atanu Roy, Bulti Nayak and
Satarupa Roy

Additional information is available at the end of the chapter

http://dx.doi.org/10.5772/intechopen.69527

Abstract

Acid mine drainage and coal fly ash both are the sibling products from man's increasing demand for power. Mining of coal from coal mines generates acid mine drainage (AMD), and burning of coal for thermal power generation produces fly ash (FA). Although both are hazardous to the ecosystem and human health, reunion of them into one would reduce their toxic effect on nature. Mining causes exposure of hidden rock materials containing an abundance of sulphide to the atmosphere. Oxidation of the metal sulphides (pyrite, as iron sulphide) within the surrounding rock and overburden generates acidity. Subsurface mining that pumps out water to prevent flooding releases acidic water to nearby areas, known as acid mine drainage. Tailings piles, mine waste rock dumps, and coal spoils contribute in AMD. Improper disposal of the fly ash contaminates the soil, water and air by leaching of the pollutants or air borne particulate matters. However filler properties and presence of macronutrients makes fly ash an excellent filler material for mine sites, and its soil-like properties help in plant growth if provided with organic carbon and nitrogen. This chapter proposes development of a green cover into AMD site after treatment of the AMD site with FA.

Keywords: acid mine drainage, fly ash, phytoreclamation, green capping, biofuel

1. Introduction

Acid mine drainage (AMD) and fly ash (FA) from coal-based thermal power plants have been recognized worldwide as major environmental problems caused by mining industries across the globe. But FA although a hazardous waste product has great beneficiation property for reclamation of other degraded landscapes like AMD sites. Mining activities accelerate the acid mine drainage by exposing sulfide-containing rocks to the weathering process. Mine tailings (byproducts leftover from mining), mine waste rock dumps, and coal spoils/overburdens all are contributing factors to acid mine drainage. Certain minerals like sulfides of iron (pyrite, FeS_2, and pyrrhotite, Fe1-XS) present in the rocks remain stable and insoluble until they come in contact with atmospheric oxygen and water. But upon exposure to water and air, they react to form sulphuric acid. This acid further causes dissolution of harmful heavy metals from the rocks. The heavy metal-loaded acid is carried off by rainwater or surface drainage to the nearby streams, rivers, or lakes, creating environmental risks. As the flowing water mixes with mine wastes and other mining leftovers, it picks up metals and other substances that pollute the water posing serious threats to ecosystems. Iron is an important metal in the AMD process. Combined with sulfate and/or sulfide, a yellow-orange solid colloquially known as yellow boy is produced which if deposited into streams can disrupt stream ecosystem. Acid mine drainage is a big problem in areas where abandoned mines are common. Processes that affect the release and transport of trace metals have to be known in order to achieve effective management and amendment of AMD and its consecutive impacts. FA from burnt coal contains certain heavy metals and nonmetals, which are otherwise hazardous if not properly disposed. The deleterious effects of AMD and FA have been studied earlier by many researchers. Tons of these wastes generated each day have certain characteristic features which can help in treating AMD. It has been reported that the acidic AMD has a pH of 2–4, along with a high sulfate concentrations (1–20 g/l). The mineralization and extremely acidic pH levels cause toxicity of these wastes, which are not only environmentally damaging but also require expensive storage, remediation, and disposal techniques. Fly ash on the other hand is alkaline in nature due to its constituent lime fraction along with other metal components. Many studies have found fly ash a very good neutralizing agent compared to chemical-neutralizing agents such as lime, slaked lime, calcium carbonate, sodium carbonate, sodium hydroxide, and magnesium oxide and hydroxide whose cost and effectiveness remain questionable [1]. Fly ash generated by coal-based thermal power plants provide ample amount of cheap resource which on successful implementation on AMD will address two main issues:

- The reduction of the AMD treatment cost.
- The effective management of FA.

As compared to the conventional processes of AMD neutralization, which have their own pros and cons, the cost involved in FA application is less. FA is freely available in close adjacency to coal mines generating AMD which even reduces the cost of transfer when mines are placed in close proximity. Millions of tons of FA generation are a matter of concern regarding proper disposal, as it poses environmental threats. Treating AMD with FA is like killing two birds with single stone. It can be effectively used to fill mine spoils and remediate acidic soils. From a pH range of 2.5–2.7, AMD can be neutralized to a pH of 7.0 with fly ash.

Increasing the pH and immobilizing the metal contaminant elements as amorphous hydroxides or oxyhydroxides, fly ash is an excellent option which contains substantial amounts of silicon dioxide (SiO_2) (both amorphous and crystalline), aluminum oxide (Al_2O_3), and calcium oxide (CaO). Liming process requires further treatments with flocculants and biological procedures for sulfate reduction, whereas soluble residues are produced when AMD is treated with FA which needs disposal. If FA is utilized in acid mine water neutralization, it can curb the additional costs of conventional chemical treatments. This will not only prove to be cost-effective but also treat and dispose both AMD and FA in an easy way. Treatment of AMD with FA not only controls acidification process but also helps in proper utilization of FA generation. Experimental procedures with FA in AMD neutralization have revealed that a predetermined mixture of acid mine water and fly ash when continuously stirred at room temperature and pressure at a specific fly ash to acid mine water ratio (AMD-FA) of 3:1 for 30 min resulted in pH increase toward making it alkaline. From a pH range of 2.5–2.7, AMD can be neutralized to a pH of 7.0 with fly ash [2]. FA has been used successfully in controlling mine drainage water through construction of a constructed wetland [3], mine backfilling [4, 5], and reclamation of abandoned open cast mine sites [6] and abandoned, underground coal mines [7]. However, the present chapter would discuss application of FA for phytoreclamation of abandoned mine sites. Mining is a temporary use of land, with the life span of an active mine ranging from few years to several decades. Before the 1970s, it was a common practice for mining companies to leave mine sites in its present state. However, now it is mandatory that mine closure would require the reclamation or rehabilitation of decommissioned mine sites, which involves returning the land and watercourses to an acceptable standard. Reclamation is the process by which degraded lands are returned to productivity and by which some measures of biotic function and productivity are restored. Various types of reclamation activities which are in use includes removing or remediating hazardous materials, reshaping the land, restoring topsoil, and planting native grasses, trees, or ground cover. Sometimes the methods used remain inadequate for the complete prevention of environmental contamination, and the mines are therefore considered to be improperly reclaimed by today's standards. Some abandoned mines are announced as "orphaned"—as the owners cannot be traced or in cases where the owner rejects to clean the mine site. In those cases, governments decide to close and rehabilitate orphaned and abandoned mines. Phytoremediation (phyto stands for plant and remediation stands for removal or correction of an evil) is a technology that uses various plants to extract, contain, degrade, or immobilize contaminants from soil and water. This technology has been receiving increasing attention for remediation of hazardous waste sites as an innovative, cost-effective, environment-friendly alternative to other treatment methods. Although phytorestoration has been successfully used for reclamation of FA ponds, phytorestoration of acid mine drainage has not been tried. The present chapter discusses the problems associated with direct phytoremediation of AMD soils. The extreme acidic pH and high sulfate concentration would not be able to provide environment for plant growth. Although few native species of plants have been found in different countries that grows in acidic soils [8], most plants studied suitable for phytoremediation would not grow in AMD sites. But it may be noteworthy to say that the AMD treated or filled with FA would present a suitable matrix for plant growth. In addition to the high concentration of plant essential nutrients such as Fe, Zn, Cu, Al, Mn, etc., which are already present in coal burned fly ash, amendment with biofertilizers and mycorrhizas can accelerate phytocover development. Soil productivity can be increased by amending soil with many types of natural materials such as animal manures, saw dust, sewage sludge, wood residues, etc. These materials stimulate the microbial activity and provides N, P, and organic carbon to the soil. The phytocover would not only have esthetic value but would also be economically profitable in long terms through cultivation of bioenergy/biofuel crops. The role of certain microorganisms (bioremediation) also has been reported which can inhabit in acidic and alkaline environment and can participate in the reclamation process [9]. They have heavy metal-accumulating properties and are capable of biotransforming them into a less toxic form. This property of heavy metal reduction by microorganisms has been widely studied and accepted by many researchers. Certain rhizobacteria (bacteria inhabiting root system of plants) are there which are heavy metal tolerant and have various plant growth-promoting features [10]. Bioaugmenting those bacteria in the site would serve additional benefits. Exclusive properties of FA as a niche for certain heavy metal-resistant bacteria would prove it to be an excellent stride towards an eco-friendly effort of better AMD beneficiation.

2. Hazards from the two by-products of mining industries

Power generation in most countries is largely dependent on thermal power plants, and they are dependent on coal, a product from mining industry. On the other hand, coal mines and other types of mines are source of acid mine drainage (AMD). Both types of by-products from the mining industry are deleterious to environment if not managed properly. Following is a discussion on the hazardous nature of coal fly ash (ash derived from coal burning) and AMD.

2.1. Ecosystem and health hazards from coal fly ash

Coal ash is the leftover waste after combustion of coal. It includes fly ash (fine powder-like particles that are driven out from the boiler of thermal power plants with the flue gases and are captured by pollution control devices) and other coarser materials that drop to the bottom of the furnace. Coal ash typically contains toxic heavy metals including arsenic, lead, mercury, cadmium, chromium, and selenium and sometimes radioactive wastes as well. The Environmental Protection Agency (EPA) has warned that living next to a coal ash and particularly unlined wet ash pond (surface impoundment) disposal site can increase risk of cancer or other diseases. Leaching of heavy metals from dump grounds to the underground water reservoir may contaminate drinking water supply with arsenic and other heavy metals. If eaten, drunk, or inhaled, the air-borne toxicants released from thermal power plants or landfills or ash ponds can cause cancer and several other diseases effecting the nervous system (cognitive deficits, developmental delays, and behavioral problems), respiratory damage of the heart and lungs, kidney disease, reproductive problems, gastrointestinal illness, and birth defects. The EPA estimates that 140 million tons of coal ashes are generated annually. Arsenic is one of the most common, and most dangerous, pollutants from coal ash. The EPA also found that living near ash ponds increases the risk of damage from cadmium, lead, and other toxic metals.

A giant spill occurred in Kingston, TN, just before Christmas 2008. More than a billion gallons of water and coal ash spilled into the neighboring river valley. EPA found arsenic at 149 times the allowable standard for drinking water in the water, as well as elevated levels of other toxic metals including lead, thallium, barium, cadmium, chromium, mercury, and nickel. Another case study of leaching hazards from FA dump occurred in Niagara (NY) Mohawk Power Corporation on Lake Erie. The drinking water wells were severely contaminated with leached lead which is a very potent neurotoxin that can harm the developing nervous system, even at low levels of exposure. Another case of contamination from use as "fill" occurred at a 216-acre golf course in Chesapeake, VA, where 1.5 million cubic yards of fly ash were recycled to give contour to the course. When groundwater at the golf course was tested, arsenic, boron, chromium, copper, lead, and vanadium were detected.

2.2. Ecosystem and health hazards from open cast, underground, abandoned, and other types of mines

2.2.1. Open cast mining

In open cast mining, material to be mined is excavated from an open pit. This type of mining is the most common type of mining for obtaining strategic minerals. This type of mining is very environmentally hazardous as strategic minerals are normally available in small amounts, which increases the amount of ore needed to be mined. Hard-rock mining causes rock exposure. When these rocks are crushed, radioactive elements, asbestos-like minerals, metallic dust, and other toxic elements get exposed. During separation, residual rock slurries, which are mixtures of pulverized rock and liquid, are produced as tailings, toxic and radioactive elements from these liquids can leak into bedrock if not properly contained.

2.2.2. Underground mining

Underground mining is very dangerous as it has the potential for tunnel collapses and land subsidence. It involves large-scale movements of waste rock and vegetation, similar to open pit mining in addition to releasing a variety of toxic compounds into the environment. As water takes on harmful concentrations of minerals and heavy metals, it becomes a contaminant. This contaminated water can pollute the region surrounding the mine and beyond. Most underground mining operations increase sedimentation in nearby rivers through their use of hydraulic pumps and suction dredges; blasting with hydraulic pumps removes ecologically valuable topsoil containing seed banks, making it difficult for vegetation to recover. Underground mine also causes deforestation leading to land erosion.

2.2.3. In situ leach (ISL) mining

ISL mining in the best type of mining over conventional mining in that the ore body is dissolved and then pumped out, leaving minimal surface disturbance and no tailings or waste rocks (World Nuclear Association, 2012). There is no formation of ore dust, and a lower water consumption is required [11]. But the use of strong acids for dissolving the ore body also dissolves metals in the host rock body. The fluids retained after the leaching process is also dangerous to the nearby ground and surface water sources as it contains elevated concentrations of heavy metals and radioactive isotopes [11]. Additionally, the acid-containing ISL mining wastewater can result in significant acidification of the surrounding environment if not neutralized by neutralizing agents.

2.2.4. Abandoned mines

Abandoned mines are areas where mining activities once occurred, but proper mine closure and reclamation did not take place or were incomplete. Ineffective enforcement on mine reclamation policy and minimal penalties for noncompliance were the main contributing factors for the large numbers of abandoned mines present today. Mines were used to be abandoned for various reasons such as reduction in the price of mining minerals, unforeseen disaster that caused bankruptcy of the mining company, national security issues in times of conflict leading to the rapid development of a number of mines without proper operations or closure requirements, political unrest causing the unscheduled halt of mining operations and small-scale illegal mining, etc. Although abandoned they still continue to release toxic products from the leftover mining residues, underground workings, open pit mine faces, waste rock piles, and tailings storage areas to the nearby lands and water bodies for a long time. The majority of the AMD actually comes from inactive or abandoned mines [12]. Until the 1970s, there was a lack of mine reclamation policies. However, nowadays most countries have adopted strong regulations according to which mining companies have to provide specific mine closure plan before their mining project is approved.

3. Physical and chemical methods of AMD treatments with their inherent advantages and disadvantages

Physical and chemical procedures of AMD treatment include prevention of AMD drainage by the following processes involved:

- Surface water diversion

In this technique, the water supply to certain acid-generating materials is reduced by the use of safeguard channel pipes, slope changes, and water proof channels.

This technique should be applied after a thorough study of hydrological and hydrogeological area, generally lack of which results in an inappropriate application of this procedure. Sometimes due to surface water diversion, certain less acid-producing areas receive less water, leading to extreme acid production. However, being a cheap technique and easily applicable, it works well in collaboration with other techniques.

- Dry covers

Dry cover application aims at achieving two different objectives:

(1) Prevention of erosion by wind and water, by stabilization of mine wastes.

(2) Pollutant releases inhibition and prevention which can be achieved by preventing oxygen and water penetration in wastes. Oxygen deficiency decreases microbial oxidation of sulfides. Different covering materials such as soil, rock, or any cement-like materials form a crust. The cover must be capable of providing prolonged protection. **Figure 1** shows a layout of a dry cover generation (reproduced from Johnson and Hallberg [13]).

Different types of materials are used for dry cover. Advantages and disadvantages of different materials used for dry covers are displayed in **Table 1**.

Sludges combined with certain inorganic covers improve the sealing properties and physical conditions of the soil. In promoting vegetation cover, it prevents erosion improving the appearance of the area.

- Sealing with clay

A convenient coating in the mine holes with clay avoids acid transfer through groundwater discharge. This reduces the water input in mining wastes. Sometimes inland water pits are made in the mines which contain the acidic water acting as a reservoir for the same. This scenario is quite common in abandoned mines, which needs remedial techniques; hence sealing is a desirable option.

3.1. Chemical procedures

Alternatively, neutralization of AMD is carried out by application of chemicals. Abatement of these unfriendly wastes is possible according to studies. Today there are many methods that are in practice on active mine sites to prevent this from happening. This includes capping of waste piles, having on-site treatment with constructs like limestone trenches [14].

The most widespread method used to mitigate acidic effluents is an active treatment process involving addition of a chemical-neutralizing agent [15, 16]. Addition of an alkaline material to AMD will raise its pH and accelerate the rate of chemical oxidation of ferrous iron.

Earlier, various chemicals including lime, slaked lime, calcium carbonate, sodium carbonate, sodium hydroxide, and magnesium oxide and hydroxide have been used to overcome the acidic fractions as a procedure. However, their cost and effectiveness remain questionable. As reported earlier by scientists, sodium hydroxide is 1.5 times effective in neutralization as compared to lime but remains a dearer option. Diminishing lime reserves and costing systems remain as issues for treatment of AMD along with the formation of ferric oxyhydroxide mantle, thereby compelling researchers to explore different options.

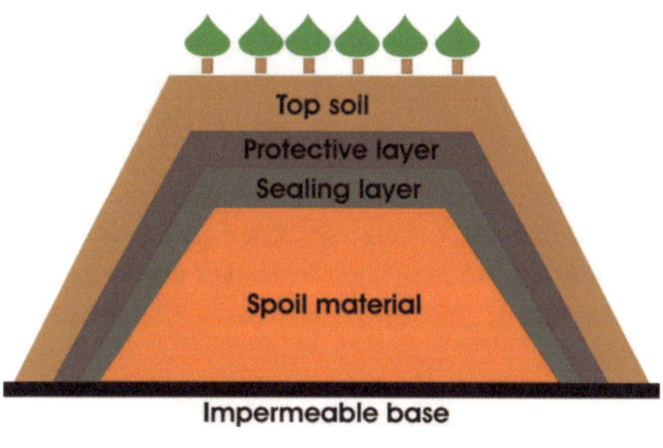

Figure 1. Layout of a dry cover for reduction of mine spoil effluents (source: Johnson and Hallberg [13]).

Material	Advantages	Disadvantages
PVC sheet	Completely impermeable Low installation costs	Capable of boring Relatively expensive
Grout	Lower price Low installation costs	Susceptible to weathering Undefined compaction/permeability rate
Clays (bentonite)	Durable Cheap and available	Undefined compaction/permeability rate Dry conditions required for application
Asphalt	Fully impermeable	Expensive material High installation costs Liable to deformation

Table 1. A comparative analysis of the advantages and disadvantages associated with different materials used for dry cover.

The studies regarding interaction of certain chemicals with tailings, and reduction in acid release and iron and other heavy metals' solubility decrease with the increase in pH, have been reported. Certain inexpensive alkaline chemicals can be added to these mining wastes, which would be beneficial in neutralization of acid pH in mining wastes. But the recurrence of acidification remains an issue.

Limestone shallow pits known as leach beds are filled with limestone chips and stone, adding alkalinity to the passing water. However, the water for treatment must be diverted toward it and should have a passage out after a waiting time. Fairly easy to construct, it seems a wetland with stone bed which is not deep. **Figure 2** shows the use of limestone for reducing the acidity of an AMD site.

Liming process involves application of a lime layer to produce an alkaline environment inhibiting bacterial growth and partial neutralization of the acid. Being a cheap alternative, the practice is highly applicable; however, the neutralizing potential becomes challenging and interrupted on deposition of iron hydroxide layer on the surface, in extreme acidic conditions. Deposition of iron on the limestone reduces its ability to neutralize the acid. Removal of this iron is necessary as its role in leaching is customary. Iron removal from solutions is achieved by precipitation with siliceous materials like furnace slag, portland cement, and fly ash. Oxidation inhibition lessens the amount of ferric ion reacting with pyrites. Various types of alkaline chemicals can be added in the mine wastes to decrease the release of acid, iron, and other metals from the waste deposit. They will neutralize the acid, and as many metals have a lower solubility at higher pH, the release of heavy metals from the mine waste would also decrease. **Figure 3** shows addition of chemicals over mine waste.

3.2. The use of bioreactors

Utilization of microorganisms for active metal precipitation is the key concept of certain bioreactors, which do not require water diversion.

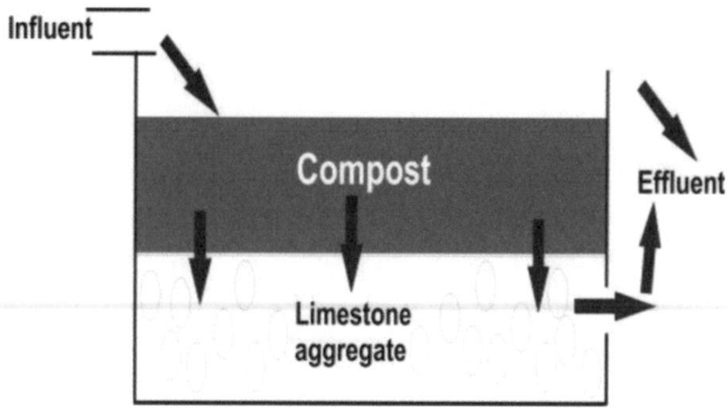

Figure 2. Schematic layout of a reducing and alkalinity producing system (RAPS), redrawn after Younger et al. [17] and Johnson and Hallberg [13].

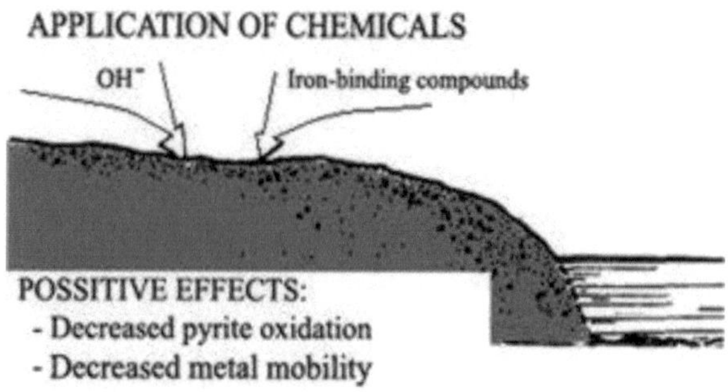

Figure 3. Treatment of solid mine waste with chemicals (redrawn from Ledin and Pedersen [9]). Application of chemicals on mine waste reduces leaching of heavy metals and decreases acid production from pyrite oxidation (source: Ledin and Pedersen [9]).

4. Treatment of acid mine drainage with fly ash: killing two birds with a single stone

Fly ash for reclamation of abandoned coal mine is an environmental approach, which is the most viable option for maximum utilization of disposed ash. Saving land requirements for ash dumping, reclamation also proves to be a better choice for gainful use of reclaimed mines.

Bulk quantity of fly ash can be utilized in lying of underground mines and open cast mine voids with preference, compared to sand [18, 19]. Backfilling of mines provides support to pillars and walls and helps reduction of void volume in underground coal mines. Studies suggest greater stability of backfilled mines than non-backfilled ones. FA works as a backfilling material effortlessly due to its pozzolanic nature which has low permeability with respect to soil or other mine wastes, thereby reducing the chance of chemical leaching. AMD generation and sinking in are reduced in FA backfilled mines, and the substituent solid residues generated remain strongly alkaline and possess pozzolanic properties. Fly ash can be an inherent candidate for mine reclamation due to its unique properties of alkaline nature, easy flowing capable as filler, good water holding capacity initiating afforestation, and a good soil ameliorant [20–22].

However, coal fly ash beneficiation in the case of AMD requires the below-listed criteria, studies of which have been acclaimed toward utilization of coal ash on active mine site. Before ash placement, certain procedural sequential steps need to be taken which would suggest successful ash placement in acid mine reclamation projects. Fly ash contains:

- Alkaline condition
- Low permeability
- Soil substitute or additive qualities
- A placement alternative

Alkaline quality generates alkaline leachates and is further used to counteract the ability of onsite materials to generate AMD. However, it is recommended that a neutralization potential and pH ranging between 7.5 and 12 are available. For proper utilization of fly ash having pozzolanic properties and low permeability, it is important that the permeability under lab conditions be equal to or less than 1.0×10^{-6} cm/s. The alkaline nature makes fly ash a good soil cover for the soil or soil substitute where it acts as a liming agent. Coal fly ash acts as an ameliorant and improves the characteristics of soil when the soil ash mixture contains a pH of 6.5–8.0. This pH is ideally the most acceptable to attain growth medium. Backfilling of pits and recontouring refused piles make utilization of FA known as ash placement. In short coal ash is beneficial in many ways, acting as a capping agent, for placements and a soil additive or enhancer.

The application and use of fly ash on-site follow certain guidelines before approval which involve detailed application for utilization of fly ash, purpose and amount of requirement, and above all details of handling and incorporation at the project site. A clear outline and indication stating the ash is physically and chemically fit for utilization are mandatory. Collection of background data and written specifications regarding hydrogeological study of the ash utilization area is required to monitor the effect of ash on surface and groundwater quality.

Before application of ash on-site, a chemical analysis on dry weight basis has to be performed (testing of metals and pH). Similarly, leachate tests by SPLP synthetic precipitation leach procedure has to be conducted for pH, sulfate, chloride, and other metals. As long as the ash is used on the mine site, periodic recertification or monitoring is required to ensure ash quality. Sometimes a mixture of bottom ash and fly ash is considered for use on-site; in such cases both the ash quality tests should be provided before utilization. A detail soil analysis report which provides data on pH, PCB, and different heavy metals and nonmetals and trace elements (micro and macro) including arsenic, cadmium, copper, chromium, lead, etc. is required when the proposed utilization of ash is in the form of a soil additive. This leads to the possibilities known for efficient plant take-up process [23].

Ash application on all site areas requires groundwater monitoring, except those where the motive of ash application is just as a soil additive, because the amount of ash addition is comparatively less. However, in all other mines, groundwater monitoring is required before and at the time of application and after ash placement. Monitoring of groundwater includes monitoring wells where there are insufficient groundwater discharge points.

Ash monitoring points should be sampled time to time so as to meet the standard mine drainage parameters along with certain metals like aluminum, cadmium, calcium, selenium, etc. A time interval for monitoring on quarterly basis is required for mine drainage parameters and annually for chlorides and additional existing metals. Moreover, some sites might require additional monitoring. On completion of the site, monitoring continues until all parameters are assessed as stable. Ash placement is done not less than and within 8 feet on top of the groundwater table of that region.

Coal ash utilization for mine reclamation takes into consideration the following special coal ash conditions:

- Grading the disposal area for a stable base.
- Coal ash disposal area should be free from invading waters at all times.
- Moisture content of the ash must be under permissible limits so as to have 90% minimum compaction.
- Deposition of coal fly ash should not be within 8 feet of any vein or seam, pit floor or any coal outcrop, or groundwater elevation.
- Chemical and leachate analysis should be conducted for coal ash on regulatory basis.
- Modified proctor or standard proctor tests should be done for all coal ash in order to check for moisture content and 95% of maximum dry density.
- Water quality monitoring on quarterly basis.
- Minimum of 4 feet should be the final ash disposal area.
- To ensure proper field compaction, field density tests have to be performed.
- Finally, all coal ash conveyed to the ash disposal areas must be spread in lifts of 2 feet or less.

The abovementioned measures not only ensure mine reclamation but also indulge in wise disposal and utilization of a waste product into a proposed, eco-restored site at no extra cost. This wise adaptation of acid mine filling also helps in afforestation of the otherwise deserted land.

In summary, FA gives following benefits for reclamation of AMD sites:

(1) The alkaline nature of FA is ideal for seal or fill material of mining sites and can prevent the generation of acid mine drainage.

(2) FA is a very good substrate for agricultural application to produce artificial soil on abandoned mine lands where native soils are not available.

(3) The acid-producing materials in the spoil can be neutralized by this alkaline amendment.

(4) A flowable fill seals and stabilizes abandoned underground mines to prevent subsidence and the production of acid mine drainage.

(5) Cost of mine site reclamation can be reduced significantly with the use of adequately available FA. This also solves storage-related issues of FA as power plants always face problems on the storage of huge quantities of FA produced as a by-product of coal burning.

Example of a successful application of using FA for treating abandoned mine was shown by Maryland Department of Natural Resources Power Plant Research Program and the Maryland Department of Environment Bureau of Mines in 1995. They jointly launched an effort to prevent acid formation from abandoned, underground coal mines. They used large volume of coal combustion products (CCPs) from power plants to create flowable grouts. They injected 100% CCP grout into the Frazee Mine, located near Friendsville, Garrett County, MD, and showed that CCP grout mixture can be beneficially used for abandoned underground coal mines to reduce acid formation as well to fill mine voids. The CCP grout mixture appears to be a high-strength, low-permeability material that can easily control mine subsidence and has no negative impact on water quality [24].

5. Posttreatment phytoreclamation of the abandoned mine

After filling the AMD site with fly ash, the next strategy would be preparing it for development of a plant cover. Although FA has many plant growth-promoting macronutrients and micronutrients, it lacks organic carbon, nitrogen, and phosphorous. So prior to plantation development, addition of various supplementations such as organic manure, biofertilizer, vermicompost, etc., is necessary [25].

5.1. Phytoremediation or phytoreclamation

Phytoreclamation or phytoremediation can be defined by three basic processes [26]:

- Plant extraction or phytoextraction: Phytoextraction is the removal of contaminants from soil or water through plant uptake and accumulation within plant tissue (either underground or aboveground biomass or both). One subtype is rhizofiltration by which hydrophytes or hydroponically grown plants use their roots in water to uptake or absorb the pollutants from water and bioconcentrate or precipitate them in the root or translocate and store them in shoots or leaves.

- Phytodegradation: Phytodegradation is the degradation or metabolism of the xenobiotic compounds by plant processes or plant-associated enzymes, bacteria, and other microflora. The endophytes present within plant tissues occasionally degrade PAH compounds that plants uptake from soil or water. Sometimes plant enzymes degrade them totally or transform (incomplete degradation) them to a less toxic state.

- Phytostabilization: Phytostabilization is the in situ immobilization or containment of the contaminants in the plant rhizosphere zone. Plants with extensive root system can reduce soil erosion to a large extent and in this procedure immobilize the heavy metals and other compounds present in soil. It is an intermittent way to prevent leaching of harmful chemicals or heavy metals from entering groundwater.

Phytoremediation is a newly evolving field of science and technology that uses plants (grasses/herbs/woody species) to clean up polluted soil, groundwater, and wastewater and through this procedure removes, contains, or renders contaminants such as heavy metals, metalloids, trace elements, organic compounds, and radioactive compounds. Phytoremediation is dependent on the extraordinary uptake capabilities or sequestration ability of plant roots. Following uptake plants can translocate and bioaccumulate them to shoots or leaves. Some pollutants also undergo phytodegradation, phytotransformation, and phytovolatilization too. Phytoremediation is environment friendly and inexpensive as it does not need expensive excavation and transport of polluted media, and this also reduces the risk of spreading the contamination and has the potential to treat sites polluted with more than one type of pollutant. Phytoreclamation of FA disposal grounds and ash ponds has been applied with documented successes [27]. It is fast becoming acceptable to the public and in most cases is less expensive than traditional treatment technologies such as incineration, bioslurry composting, etc.

5.2. Production of a phytocover

The initial phytocover development is generally made by growing a mat of grasses and legumes [28, 29]. Grass-legume cover by the native species (and sometimes invasive species) has become the most efficient method of choice as they can readily colonize the area and develop a thick vegetation mat within a short period of time. In the meantime, they enhance fertility of the soil (FA amended with nutrients), prevent erosion and emission of air pollutants, and also phytosequester the heavy metals. This paves way for long-term management and gradual restoration of the site. However, prior to consideration of phytoremediation, factors like studying characteristics of the substrate, properties of the plants and residual microbes, and effect of the phytocover that it would have on the nearby ecosystem should be studied. Natural succession on FA basins has been studied in different countries. Grasses are the primary colonizers in degraded soils as they are better tolerant to drought and other environmental stresses (e.g., xenobiotic compound load) and require fewer nutrients and proliferate rapidly. The fibrous nature of the grass roots is effective in slowing down erosion and preserving soil moisture. Their soil-forming characters eventually produce a layer of organic soil upon which other leguminous crops grow and add nitrogen. Grasses and legumes play great roles for formation of a nutrient-rich top soil. Native species of grasses have ready adaptability with the prevailing environmental conditions, increase soil fertility, have faster establishment, and provide a sustainable microclimate for establishment of commercially useful species. This gives economical return from the rehabilitation programs. The initial cover must allow the development of diverse self-sustaining plant communities. Recently, Pandey et al. [29, 30] and Żołnierz et al. [31] have enumerated various naturally growing species on FA deposits during initial colonization and 11 years after vegetation establishment. Some examples of naturally growing species are *Saccharum munja* [32], non-nodulated species *Cassia siamea Lamk*, nodulated species of chickpea [33], and *Pteris vittata*, a fern [34]. *S. munja* was termed as an "ecological engineer" due to properties such as firm ash-soil binding capability and stabilizes the ash dump surface. According to Pandey [35], *Ricinus communis* L., naturally growing grass species on FA land fill sites, is a suitable plant species for revegetation of degraded lands like mine sites or ash dump sites in tropical and subtropical climates. In addition to acting as carbon sequestrator, it has great commercial value as it is an industrially valuable castor oil yielding crop. It provides ecosystem benefits like carbon sequestration, heavy metal uptake and accumulation, phytodegradation of organochlorine pesticides, substrate quality enhancement, biodiversity conservation, and feedstock for biofuel production [36]. Ferns like *P. vittata* L., *Ampelopteris prolifera* (Retz.) Copel., *Diplazium esculentum* (Retz.) Sw. and *Thelypteris dentata* (Forsk.), etc. also grow well on FA without showing heavy metal-related toxicity symptom. Kumari et al. [37] reported that the fern *T. dentata* is also suitable for phytoremediation and can be used efficiently to revegetate FA landfills. Some woody tree species like *Leucaena leucocephala, Dendrocalamus strictus,* and *Eucalyptus sp.* have the potential power for revegetation of FA disposal sites. These tree species have huge economic importance. Some timber and plywood-yielding plants that have been used successfully in studies are *Shorea robusta, Tectona grandis, D. sissoo, Bombax ceiba, Populus euphratica,* and *Eucalyptus tereticornis* [38]. Examples of some fuel wood tree species recommended by some workers recommended for phytorestoration job are *Acacia auriculiformis, Albizia lebbeck, Acacia nilotica, Casuarina equisetifolia, Cassia siamea, Dalbergia sissoo, and Prosopis juliflora.* In addition to their economic importance, they have nitrogen-fixing properties and excellent growth characteristics in nutrient-limited conditions. Examples of plywood-producing tree species which also can be used to act as fuels are *Eucalyptus hybrid, Eucalyptus globules, Melia azedarach, Populus deltoids, Tamarindus indica,* and *Syzygium cumini* [39]. In a study by Carlson and Adriano [40], it was depicted that a new ecosystem can been created on alkaline and acidic FA dump sites with *Platanus occidentalis* (sycamore) and *Liquidambar styraciflua* (sweetgum), which are important timber trees.

Trees contribute greatly to the formation of self-sustaining nutrient-rich top soil formation through biological nitrogen fixation, maintenance of soil organic matter, and increasing water infiltration and storage and reduce loss of nutrients by erosion and leaching. They also improve soil physical structure through reduction of soil acidity and improvement of soil biological activity. Through plant litters and root exudates, nutrient cycle between soil and plants is established. Nitrogen is a major limiting factor of mine spoils, and regular addition of nitrogen-rich fertilizer may be required to maintain vegetation on mine sites. Introduction of legumes and other nitrogen-fixing species may solve this problem. Formation of a productive forest ecosystem from primary or secondary plant community serves as the habitat for biotic communities, establishes its own food-chain trophic levels, and allows biogeochemical cycles. Most tedious process of fly ash dump reclamation or mine site reclamation is the establishment of the initial colonizing grass species. After the initial establishment of grass cover, strategies such as additions of organic amendments and spreading of grass legume fodder seeds, forage legumes, tuft, or hardy grasses are an efficient choice for development of a forest community.

The grass legume cover helps in the development of a vegetation mat in a small period of time when compared to the tree species. A reclaimed site can then be used for different purposes like agricultural use or agro-forestry or silviculture. Finally, evaluation must be carried out to determine the success of the reclamation and monitor the groundwater quality. It is to also necessary to determine how closely the reclaimed site functions, in comparison to similar undisturbed sites, as an ecosystem. Reclamation of abandoned mine land is a very complex process. Most researchers agree that reclamation success must be measured by several parameters since no individual parameter provides sufficient information for ecosystem reclamation [41].

6. Conclusion

Ecological restoration of abandoned mine spoil sites has become an important part of the sustainable development strategy in many countries. Long-term mine spoil reclamation requires the establishment of a stable nutrient cycle from plant growth and microbial processes. Application of FA would help formation of soil over mine spoil by neutralizing the acidity and adding other important properties of soil. Growth of grasses and legumes over this would further help in the formation of top soil and by stable nutrient cycling would return the land to its pristine condition that can survive as a self-sustaining ecosystem. A successful green cap on the mine site would control pollution, stabilize the site, and give visual and esthetic pleasure.

Author details

Madhumita Roy*, Roopali Roychowdhury, Pritam Mukherjee, Atanu Roy, Bulti Nayak and Satarupa Roy

*Address all correspondence to: biotech4job@gmail.com

Department of Biotechnology, Techno India University, Salt Lake, Kolkata, West Bengal, India

References

[1] Skousen J, Ziemkiewicz P, Yang JE. Use of coal combustion by-products in mine reclamation: Review of case studies in the USA. Geosystem Engineering. 2012;**15**(1):71-83

[2] Vadapalli VR, Klink MJ, Etchebers O, Petrik LF, Gitari W, White RA, Iwuoha E. Neutralization of acid mine drainage using fly ash, and strength development of the resulting solid residues. South African Journal of Science. 2008;**104**(7-8):317-322

[3] Machemer SD, Wildeman TR. Adsorption compared with sulfide precipitation as metal removal processes from acid mine drainage in a constructed wetland. Journal of Contaminant Hydrology. 1992;**9**(1-2):115-131

[4] Williams D, Ramlackhan M, Spriggs D. Report on potential for backfilling bord and pillar voids using fly ash slurry. (2010). www.dnrm.qld.gov.au/__data/assets/pdf_file/0007/262663/collingwood-park-report-appendix-c.pdf

[5] Horiuchi S, Kawaguchi M, Yasuhara K. Effective use of fly ash slurry as fill material. Journal of Hazardous Materials. 2000;**76**(2-3):301-337

[6] Hornberger RJ, Dalberto AD, Menghini MJ et al. Coal Ash Beneficial Use at Mine Sites in Pennsylvania. World of Coal Ash (WOCA); Proceedings 2005, World of Coal Ash (WOCA), Lexington, KY, April 11-15

[7] Petzrick P, Rafalko LG, Lyons C, Chiang SH (Ed.) (1996). An overview of the western Maryland coal combustion by-products/acid mine drainage initiative, Part 1 of 3. United States: Pittsburgh Coal Conference

[8] Abreua MM, Tavaresa MT, Batista MJ. Potential use of Erica andevalensis and Erica australis in phytoremediation of sulphide mine environments: São Domingos, Portugal. Journal of Geochemical Exploration. 2008;**96**:210-222

[9] Ledin M, Pedersen K. The environmental impact of mine wastes – Roles of microorganisms and their significance in treatment of mine wastes. Earth-Science Reviews. 1996;**41**(1-2):67-108

[10] Rau N, Mishra V, Sharma M, Das MK, Ahaluwalia K, Sharma RS. Evaluation of functional diversity in rhizobacterial taxa of a wild grass (*Saccharum ravennae*) colonizing abandoned fly ash dumps in Delhi urban ecosystem. Soil Biology & Biochemistry. 2009;**41**:813-821

[11] Environment Protection Authority (EPA). Annual Report on the administration of the Radiation Protection and Control Act 1982. 2005–2006. www.epa.sa.gov.au/files/8366_radiation0506.pdf

[12] Johnson DB. Chemical and microbiological characteristics of mineral spoils and drainage waters at abandoned coal and metal mines. Water, Air, & Soil Pollution: Focus, 2003;**3**: 47-66. doi:10.1023/A:1022107520836

[13] Johnson DB, Hallberg KB. Acid mine drainage remediation options: A review. Science of the Total Environment. 2005;**338**:3-14

[14] Skousen J, Foreman J. Water management techniques for acid mine drainage control. Green Lands. 2000;**30**:44-53

[15] Coulton R, Bullen C, Dolan J, Hallet C, Wright J, Marsden C. Wheal Jane mine water active treatment plant-design, construction and operation. Land Contam Reclam 2003;**11**:245-252

[16] Coulton R, Bullen C, Hallet C. The design and optimization of active mine water treatment plants. Land Contam Reclam 2003;**11**:273-279

[17] Younger PL, Jayaweera A, Elliot A, Wood R, Amos P, Daugherty AJ, et al. Passive treatment of acidic mine waters in subsurface flow systems: Exploring RAPS and permeable reactive barriers. Land Contamination & Reclamation. 2003;**11**:127-135

[18] Pradhan A, Deshmukh JP. Utilization of fly ash for treatment of acid mine water. Journal of Environmental Research And Development. 2008;**3**(1):137-142

[19] Reynolds K, Petrik L. The use of fly ash for the control and treatment of acid mine drainage, in: Proceedings of world of coal ash symposium 2005, Lexington, USA

[20] Gitari MW, Petrik LF, Etchebers O, Key DL, Iwuoha E, Okujeni C. Treatment of acid mine drainage with fly ash: Removal of major contaminants and trace elements. Journal of Environmental Science and Health Part A. 2006;**41**(8):1729-1747

[21] Gitari WM, Petrik LF, Etchebers O, Key DL, Okujeni C. Utilization of fly ash for treatment of coal mines wastewater: Solubility controls on major inorganic contaminants. Fuel. 2008;**87**(12):2450-2462

[22] Gitari WM, Somerset VS, Petrik LF, Key D, Iwuoha E, Okujeni C. Treatment of acid mine drainage with fly ash: Removal of major, minor elements, SO_4 and utilization of the solid residues for wastewater treatment. In: International Ash Utilization Symposium, Center for Applied Energy Research, University of Kentucky. 2003. pp. 1-23

[23] Luptakova A, Ubaldini S, Macingova E, Fornari P, Giuliano V. Application of physical-chemical and biological-chemical methods for heavy metals removal from acid mine drainage. Journal of Biotechnology. 2010;**150**:252-253

[24] Guynn RL, Rafalco LG, Petzrick P. Use of a CCP Grout to Reduce the Formation of Acid Mine Drainage: 10-Year Update on the Winding Ridge Project. Presented at the World of Coal Ash Conference Proceedings, Lexington, Kentucky, 2007. http://www.flyash.info/2007/52guynn.pdf

[25] Juwarkarr AA, Jambhulkar HP. Restoration of fly ash dump through biological interventions. Environmental Monitoring and Assessment. 2008;**139**(1):355-365

[26] Cunningham SD, Lee CR. Phytoremediation: Plant-based remediation of contaminated soils and sediments. Bioremediation: Science and applications. Soil Science Society of America. 1995;**43**:145-156

[27] Mpofu K, Mushiri S, Mushiri K. An investigation into the effectiveness of coal ash in acid mine drainage (AMD) abatement. A case study of iron duke mine. International Researcher. 2013;**2**(3):49-65

[28] Prasad MNV. Engineered phyto-covers as natural caps for containment of hazardous mine and municipal solid waste dump sites–possible energy sources. In: Ozturk M et al (eds) Phytoremediation for Green Energy. Springer Netherlands. 2014. pp. 55-68

[29] Pandey VC, Prakash P, Bajpai O, Kumar A, Singh N. Phytodiversity on fly ash deposits: Evaluation of naturally colonized species for sustainable phytorestoration. Environmental Science and Pollution Research. 2014;**22**(4):2776-2787

[30] Pandey VC, Bajpai O, Pandey DN, Singh N. *Saccharum spontaneum*: An underutilized tall grass for revegetation and restoration programs. Genetic Resources and Crop Evolution. 2015;**62**:443-450

[31] Żołnierz L, Weber J, Gilewska M, Strączyńska S, Pruchniewicz D. The spontaneous development of understory vegetation on reclaimed and afforested post-mine excavation filled with fly ash. Catena. 2016;**136**:84-90

[32] Pandey VC, Singh K, Singh RP, Singh B. Naturally growing *Saccharum munja* on the fly ash lagoons: A potential ecological engineer for the revegetation and stabilization. Ecological Engineering. 2012;**40**:95-99

[33] Pandey VC, Singh JS, Kumar A, Tewari DD. Accumulation of heavy metals by chickpea grown in FA treated soil: Effect on antioxidants. Clean-Soil Air Water. 2010;**38**:1116-1123

[34] Srivastava M, Ma LQ, Singh N, Singh S. Antioxidant responses of hyper-accumulator and sensitive fern species to arsenic. Journal of Experimental Botany. 2005;**56**:1335-1342

[35] Pandey VC. Suitability of Ricinus communis L. cultivation for phytoremediation of fly ash disposal sites. Ecological engineering. 2013;**57**:336-341

[36] Rissato SR, Galhiane MS, Fernandes JR, Gerenutti M, Gomes HM, Ribeiro R, Almeida MV. Evaluation of Ricinus communis L. for the phytoremediation of polluted soil with organochlorine pesticides. BioMed Research International. vol. 2015, Article ID 549863, doi:10.1155/2015/549863

[37] Kumari R, Kaur I, Bhatnagar AK. Enhancing soil health and productivity of *Lycopersicon esculentum* Mill. using *Sargassum johnstonii* Setchell and Gardner as a soil conditioner and fertilizer. Journal of Applied Phycology. 2013;**25**(4):1225-1235

[38] Ram LC, Jha SK, Tripathi RC, Masto RE, Selvi VA. Remediation of fly ash landfills through plantation. Remediation Journal. 2008;**18**(4):71-90

[39] Pandey VC, Abhilash PC, Singh N. The Indian perspective of utilizing fly ash in phytoremediation, phytomanagement and biomass production. Journal of Environmental Management. 2009;**90**:2943-2958

[40] Carlson CL, Adriano DC. Growth and elemental content of two tree species growing on abandoned coal fly ash basins. Journal of Environment Quality. 1991;**20**:581-587

[41] Sheoran V, Sheoran AS, Poonia P. Soil reclamation of abandoned mine land by revegetation: A review. International Journal of Soil, Sediment and Water. 2010;**3**(2):13

Chapter 7

Challenges in Recovery of Valuable and Hazardous Elements from Bulk Fly Ash and Options for Increasing Fly Ash Utilization

Ajit Behera and Soumya Sanjeeb Mohapatra

Additional information is available at the end of the chapter

http://dx.doi.org/10.5772/intechopen.69469

Abstract

Beneficiation of fly ash should require for ensuring the removal of reactive elements to reduce the effect of hazardous impact on our atmosphere and can fill the demand for resources such as metals and rare earths. In this chapter, we concentrate to describe the responsible factors involve in fly ash beneficiation that has a great contribution to our environment. The purpose of the current study is to know the recovery of different minerals; maximum removal of the contaminant, reactivity and neutralization capacity of acid mine drainage (AMD) with fly ash and development of the cost-effective method of disposal of fly ash are achieved. Different beneficiation techniques of fly ash and utilization of fly ash are explained.

Keywords: fly ash, separation technique, fly ash utilization, acid mine drainage

1. Introduction

Fly ash is the finest form of the residue generate from coal combustion in different power plant and industries. The main cause of formation of residue is the conversion of mineral matter present in coal particles during combustion. As a consequence of coal-based power generation industries, a large amount of fly ash produced in thermal power plants create various disposal-related problems. Fly ash continues to be treated as a waste product, which contributes nearly 85% of the fly ash generated in India. It is found that in Odisha specific ash generation is 7–9 MT/day/MW and annual ash generation is about 25 MTPA in 2014–2015 (**Figure 1**), and various areas in which fly ash can be utilized in Odisha are presented in **Figure 2**. In India, ash generation likely to increase to about 180 MTPA after commissioning of proposed power plants, which required 120,000 hectares of land for disposal [1–3]. A large number of power

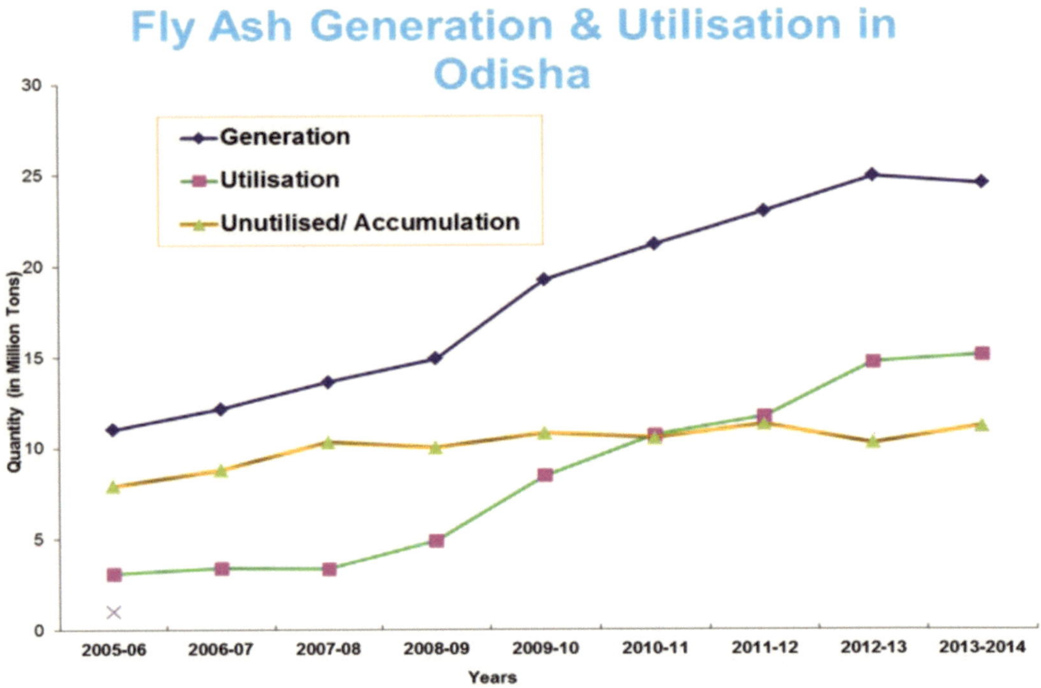

Figure 1. Fly ash generation and utilization in Odisha, India.

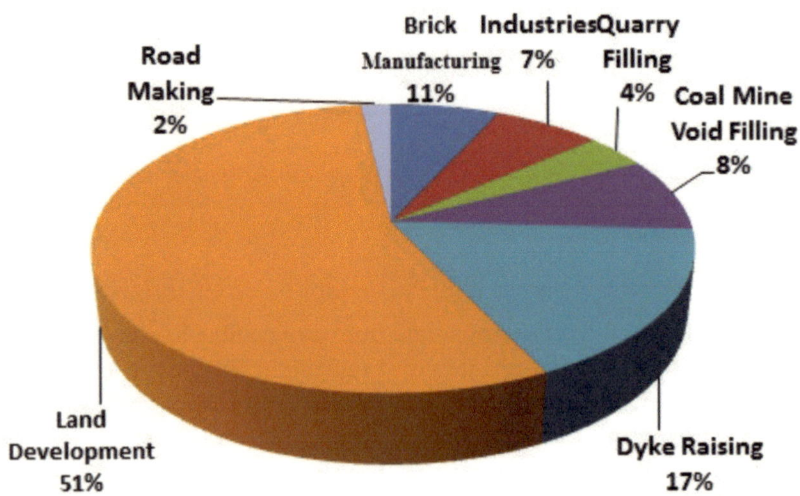

Figure 2. Status of utilization of fly ash in Odisha (2014–2015).

plants in India utilize the wet fly ash that results in degradation of the pozzolanic characteristics of the ash, an essential ingredient for several ash-based products. In India, Fly Ash Mission was set up in 1994 for bulk utilization of ash. Fly ash notification came out in 1999 and amended in 2009 for enhancing fly ash utilization in various sectors [4]. Fly ashes can be broadly classified on the basis of cementitious properties (Class C: SiO_2 + Al_2O_3 + Fe_2O_3 minimum 50 wt%) and on the basis of pozzolanic activity (Class F: the sum of the aforementioned oxides exceeds 70 wt%) [5]. Fly ash contains heavy metals (such as Fe, Mn, Al), different elements (such as Zn, Co, Pb, Cr, Cu), toxic/reactive elements (anions like SO_4), radioactive elements and hazardous elements, which give an increasing trend of negative impact to our environment year after year. Thus, in the coming years, fly ash recovery would be a major challenge. Recovering coal ash from the hazardous materials extends the life of available landfill space and reduces the negative impact on the environment. Recovered ash can fulfil the demand for resources such as metals and rare earths. Ash recovery and processing requires very less energy and produce significantly lower emissions than that of traditional mining and processing of natural mineral resources. Processing of fly ash by different techniques to make it more suitable for a specific utilization is known as beneficiation of fly ash.

Fly ash consists of inorganic, incombustible matter present in the coal that has been fused during combustion into a glassy, amorphous structure. Coal ash consists mainly of aluminosilicate glass matrix in addition to crystalline mullite and quartz. The glass phase in coal ash has been attributed to the formation of soluble salts of As, Na, Mg, K, Sr and B on the surface of coal ash spheres, which aids in the reduction in concentration of these elements. Fly ash must meet required ASTM specification before use in concrete. The loss on ignition (LOI) of fly ash is an important tool to measure the presence of unburned carbon. ASTM C618-89a gives the standard specifications for the use of fly ash in Portland cement with LOI limit of 6% [6].

Disposal of fly ash has several problems, including (1) huge requirement of water (about 36 million litres of water each day for a 1000-MW coal-fired power plant using coal with 40% ash) for draining towards the ponds, (2) creation of large wasteland and (3) health and environmental concerns associated with the likelihood of leaching of heavy metals as well as air-borne ash.

2. Environmental impact of fly ash

Fly ash utilization generates the benefit of resource conservation, reduction of waste, avoided emissions with more durable infrastructure and restoration of industrial sites. Major areas of fly ash utilization are as follows: (1) making different types of bricks/blocks for building construction, (2) manufacture of cement and additives in concrete, (3) road construction and (4) embankment, backfill, land development, etc. Land-filled ash also used in same beneficial applications as that of direct ash from the power utility. In other sides, the fly ash contains traces of heavy metals and toxic metals, which are generally below the level that can cause any impact on the environment or public during its normal use. Some elements of fly ash are radioactive. The radioactivity level of fly ash is 1/10th to 1/20th that of the level that causes any harm to environment. This has been tested and certified by Department of Atomic Energy, Government of India. Fly ash particles are generally spherical in shape and range in size from 0.5 to 300 μm, which can cause pulmonary disease. Fly ash able to accumulate toxic elements (Pb, Cr, B, Mo, As and Se) at higher temperatures and here it is considered as an environmental hazard Also due to some elements present in fly ash cause acid drainage from fly ash mine when in a larger period of contact with rain water or moisture. Therefore, the increased acidity has a range of negative effects calculated from the severity of the pH change. Acidic flow towards river system causes inhospitable to aquatic life that results in periodic fish killing, leading to ecological destruction in watersheds. The site requires expensive active maintenance to prevent the acidic drainage from contaminating the drinking water of nearby communities. Therefore, ash recovery is necessary for social, economic and environmental benefits. The recovery can fill demand for natural resources such as metals, rare earths, lime, sand and aggregate and can increase the life of available landfill space. Nowadays, several types of innovations implement for transforming today's coal ash landfills into accessible, resource-rich stores of needed materials.

3. Composition of fly ash

Fly ash constituents depend upon the source industries specific coal bed but may include substantial amounts of oxide elements such as SiO_2 both in amorphous and crystalline, Al_2O_3 and CaO and trace concentrations of elements (up to hundreds ppm) such as arsenic, beryllium, boron, cadmium, chromium, cobalt, lead, manganese, mercury, molybdenum, selenium, strontium, thallium, vanadium, dioxins and Polycyclic aromatic hydrocarbon (PAH) compounds (**Figure 3**). Besides the chemical composition, loss on ignition (LOI) reduction will be a requirement and that mentioned for fly ash to meet the specifications most notably the carbon content. Fly ash is a potential source for mineral carbonation by proper chemical co-ordination to capture atmospheric CO_2. The carbonated material can find application as a substrate and a soil conditioner for agriculture. In other cases, when some other types of fuel used instead of fly ash to

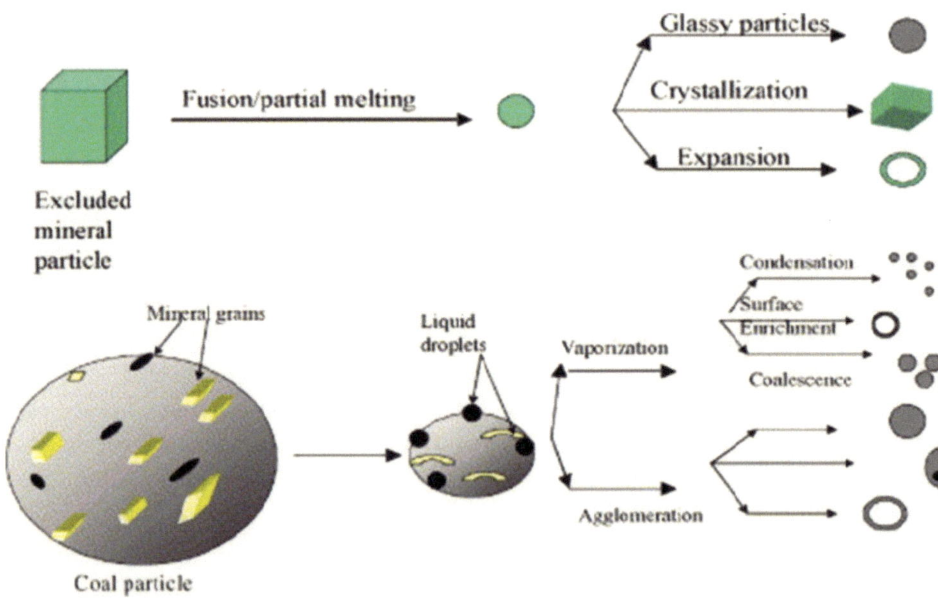

Figure 3. Entrapments of minerals in fly ash during combustion.

produce electricity in an incinerator, this kind of ash may contain higher levels of contaminants than coal ash that is often classified as hazardous waste. Petrologic analyses of fly ash from different industries were conducted using a high-resolution optical microscope. It helps to detect the percentage and aggregation of materials in fly ash. It is found that the unburned carbon is not uniform and there are three carbon types identified: (i) inertinite particles (semifusinite/maceral), (ii) isotropic coke and (iii) anisotropic coke.

4. Extraction techniques

Physical beneficiation processes enhance the utility of fly ash and are based on differences in size, density, magnetic properties, electrostatic properties and other physical properties of the particles. Chemical separation processes are also achieving a success route for beneficiation of the fly ash. Various types of physical and chemical beneficiation techniques of fly ash are given in the next section.

4.1. Sieving

Sieve analysis involves differentiating particle size distributions of fly ash as well as the carbon and minerals in fly ash. In sieve analysis, a stack of sieve with wire mesh screen attached with a column. The mesh size gradually reduces from top to bottom. Generally, a series of four sieve stacks (sizes 90, 75, 45 and 25 μ) and a bottom pan (receiver) were used for sieve analysis. A 100 g sample was put into the top sieve. The column was placed in a mechanical shaker. After the shaking, the material on each sieve is weighed. The weight of the sample of each sieve was then divided by the total weight to give a percentage retained on each sieve [7].

4.2. Air classifier

In this practice, the fly ash feed in the classifier with the main air flow and aerodynamically classifies the material (**Figure 4**). In this process, air pollution material having less in size can be recovered. There is a classifying wheel, in which, a frequency converter equipped to control the wheel speed. The size of the particle can be adjusted during operation by controlling the speed of classifying wheel. Fine particles are transported by the rotor blades along with the classifying air. These fine sizes are then discharged and collected in a suitable filter, whereas the coarse particles are rejected. The coarse material is rinsed intensively in a spiral flow taken from the secondary air stream to remove any last fine powder and are discharged via the coarse material discharge. The ultrafine fly ash produced can utilize in high-strength concrete [8].

4.3. Density separation

Different minerals in the fly ash have different density. If a mixture of minerals with different densities can be put in a liquid with an intermediate density, the grains with densities less than that of the liquid will float and grains with densities greater than the liquid will sink. The fly ash by flotation on water produces cenospheres and fly ash particles can be considered to be cenospheres. Cenospheres are the hollow glass spheres having relative density <1. The fly ash

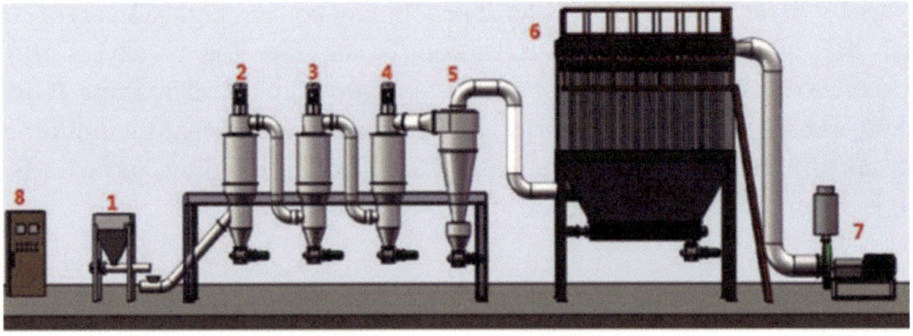

Figure 4. Arrangement of air classifier.

floaters are generally larger (40–250 μm) than the heavier fraction. With unique properties, cenospheres establish fly ash utilization in the refractory and polymer industries. Sphericity of cenospheres, controlled particle size and low density allow the manufacture of polymers with improved properties and/or reduced cost [9]. The applications of cenospheres in thermoplastics have been limited by the high-shear conditions in extrusion and injection moulding. In comparison with other irregularly shaped fly ash fillers, higher volume loadings can be achieved without significantly affecting the rheology and are used in Polyvinyl chloride (PVC), under-body sealants, shoe soles, body fillers, oil well cements, polyurethane and polyester-simulated wooden furniture and fibreglass. Cenospheres are applied in oil well cements that improve the pumpability (sphericity) and maintain the consistency of the mix by maintaining the other components in suspension and limiting the segregation (**Figure 5**).

4.4. Magnetic separation

Fly ash contains both magnetic and nonmagnetic mineralogical composition and distribution of different minerals in the particle. It is very easy to separate the magnetic and nonmagnetic particles using magnetic belt separator shown in **Figure 6**. Powdered minerals put on the belt by a feeder and the belt rotates along with the powder. When the powder reached at the magnetic roll, ferromagnetic and strongly paramagnetic particles attract towards the roll and fall on separate pan. There is a use of aerodynamic forces in the configuration to augment the magnetic force of separation. Ferromagnetic and strongly paramagnetic can be separated by adjusting the induced magnetization in the ash. For a certain set of separation conditions in the magnetic separation process, the yields of magnetic fractions were directly proportional to the saturation magnetization of the ferromagnetic components of the ash [10]. Generally, more than 50% recovery of concentrated ferrous materials can be achieved using a magnetic separator.

4.5. Electrostatic separation

Elements that chemically and mineralogically differ present in the fly ash can be recovered using electrostatic precipitator [11]. In the electrostatic separation process, particle size is

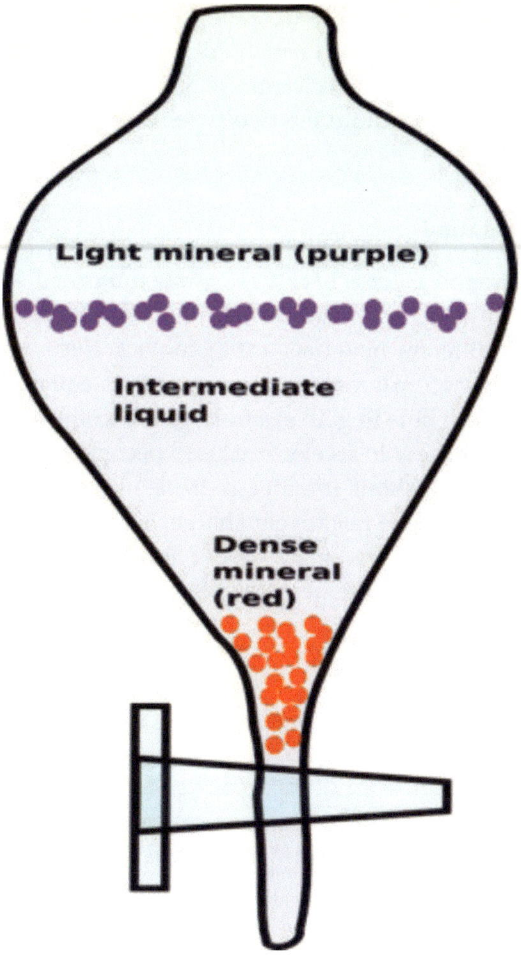

Figure 5. Density separation chamber.

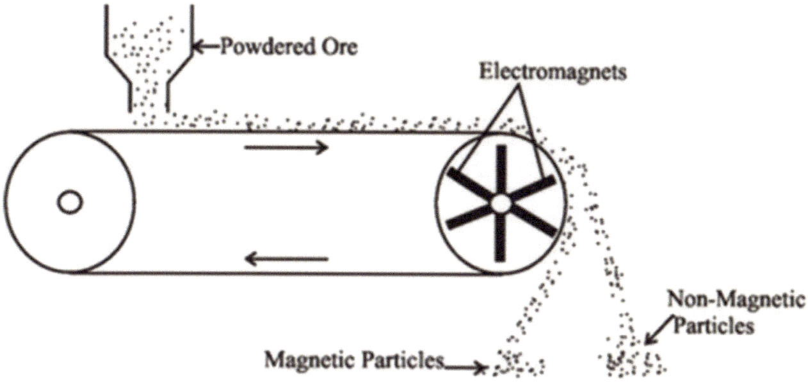

Figure 6. Magnetic separator.

approximately less than 45 μm fraction. Electrostatic parametric selection and careful blending of the various fractions recovered in different holding tank (**Figure 7**). Electrostatic charge induced on the surface plays role when selection is based on electrical susceptibility. During separation of Class F ash, it can separate the two types of glass such as alumina silicate and calcium alumina silicate.

4.6. Triboelectrostatic separation

The triboelectrostatic separation process used to separate unburned carbon and mineral particles. In this process, the organic and mineral particles are charged with opposite polarity (**Figure 8**). When the two different materials are in contact, electrons move until the energy level of the electrons in the two materials at the interface is equalized. The material with a higher affinity for electrons can able to gain electrons and become negatively charge, whereas the material with the lower affinity loses electrons and become positively charge. On contact with a tribocharger, unburned carbons (organic particles) become positively charge and the mineral (inorganic particles) become negatively charge. The differential charging of different particles in the triboelectrostatic method makes it possible to use a static high voltage separator. Carbon particles are attracted to the negative plate and mineral particles are attracted to the positive plate [12].

4.7. Column flotation

Column flotation cells shown in **Figure 9** do not use any mechanical agitation/impellers. Mixing can be done by the turbulence of rising bubbles. In this method, the ground ore is mixed with water to form slurry. The desired mineral becomes hydrophobic in contact with surfactant or collector chemical. But some mineral surfaces are naturally hydrophobic, requiring a trace amount of collector. The collector depends on the nature of the mineral to be recovered. As an example, the collector sodium-ethyl-xanthate ($CH_3CH_2OCS_2Na$) added in the flotation of PbS to separate it from ZnS. Produced slurry/pulp of hydrophilic and

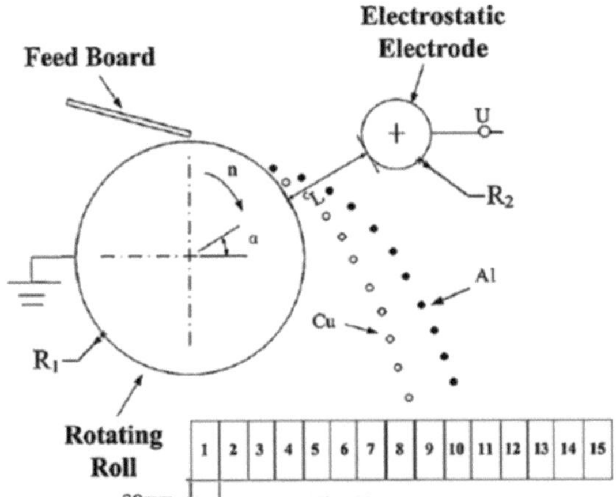

Figure 7. Electrostatic separator.

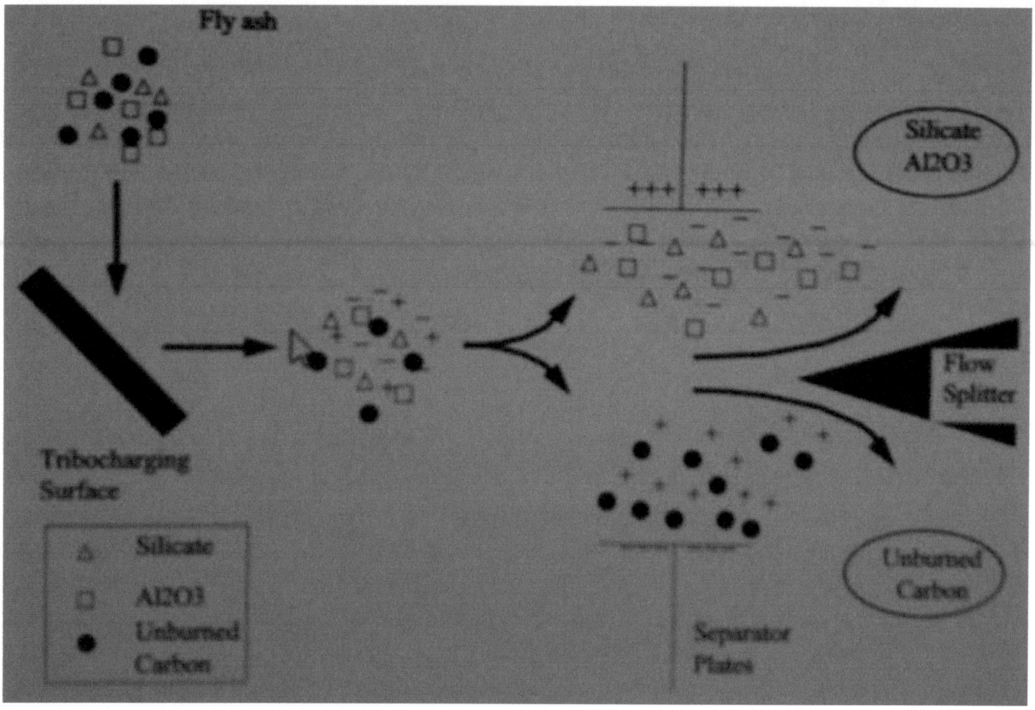

Figure 8. Principles of triboelectrostatic separation.

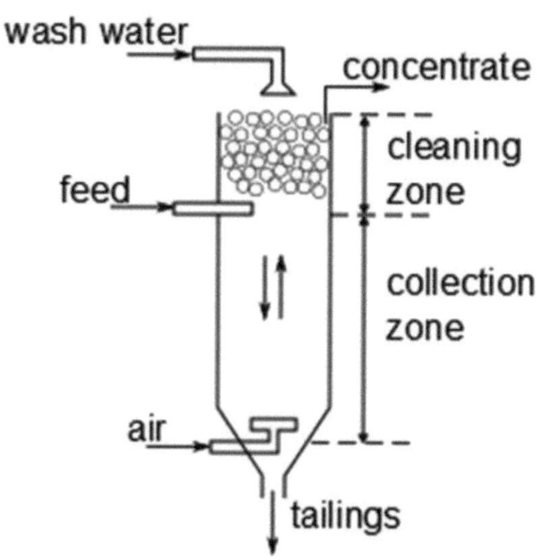

Figure 9. Column flotation cells.

hydrophobic particles introduced in flotation tank that is equipped with bubble riser. The hydrophobic particles attached with the carrier air bubbles rises to the surface and form froth. The froth is collected from the cell that producing a concentrate of the target mineral [13].

4.8. Ultrasonic column agglomeration

Ultrasonic column agglomeration consists of a column, variable speed electrical motor, a solvent recovery tank and stainless steel screen (**Figure 10**). This set-up consists of pre-conditioning treatment in the mix tank and treatment in the column [14]. Ultrasonic wave applied to the slurry mixture in the mix tank and column using a transducer. Ultrasonic wave produced in a frequency range of 0–1 MHz. Slurry of cyclohexane and fly ash was prepared with 5:1 weight ratio. This pre-conditioned slurry exposed to a ultrasonic wave of 0.5 MHz. Then the slurry pumped into the column with the column ultrasonic wave frequency. The agitation speed was controlled. The carbon concentrate was collected on the screen (60-mesh) and air dried. All the carbon recoveries were calculated on a total weight carbon basis.

4.9. Fluidized bed separation (FBA)

Fluidized bed ash formed by fluidized bed combustion in a holding vessel. The quantity of solid particulate placed under appropriate conditions to form a fluid (mixture of solid/fluid) (**Figure 11**). Generally, this solid/fluid mixture is introduced by the pressurized fluid through the particulate medium or by pumping of mixture using fluid type technologies. Now the solid/fluid mixture acts as a normal fluid and able to flow freely under gravity. The fluidized bed is one of the best-known contacting methods. The coal burn with dolomite or limestone in the fluidized bed burning the unburned carbon and releasing SO_2 from exhaust gases. Fly ash collected from the air-controlled equipment and dumped with bottom ash. This blend of this mixture is called FBA [15]. The calcium percentage in FBA is more than in fly ash. FBA has

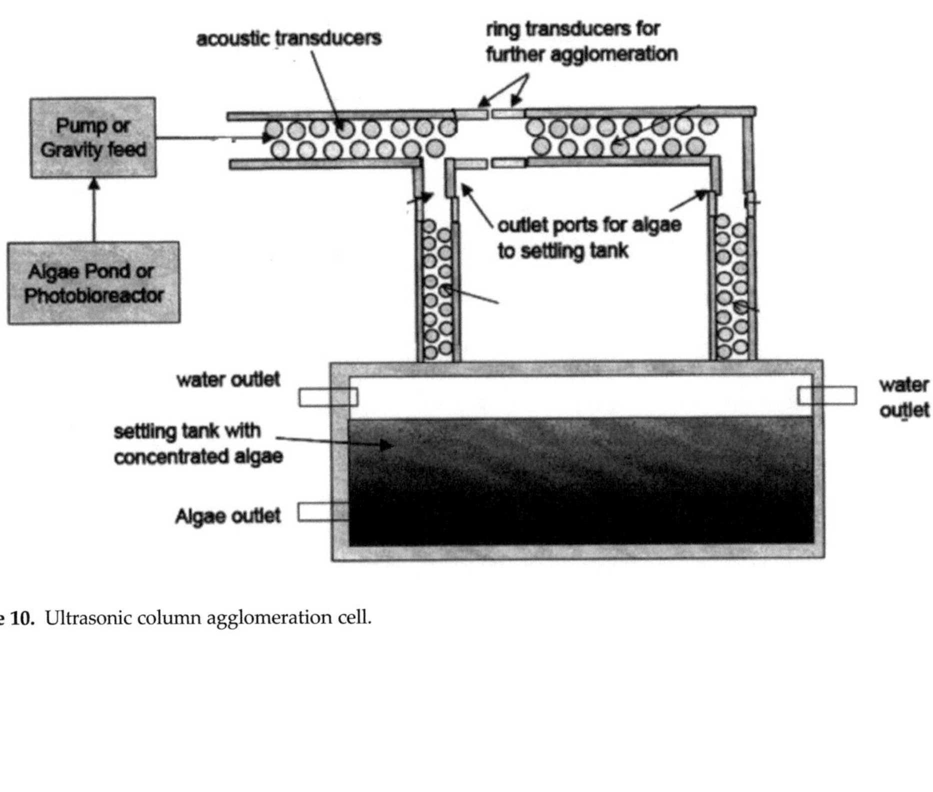

Figure 10. Ultrasonic column agglomeration cell.

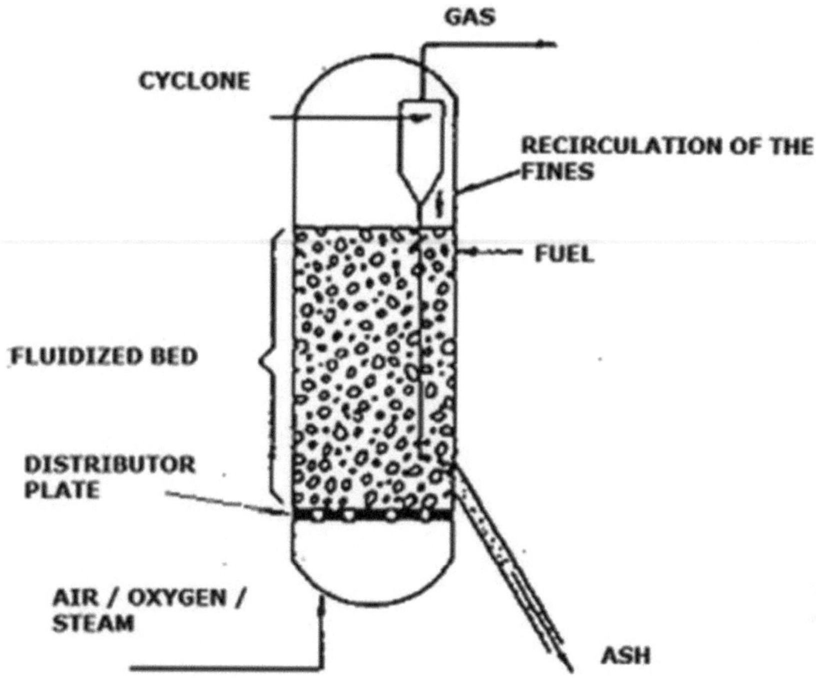

Figure 11. Fluidized bed separation chamber.

been applied to agricultural and brine contaminated lands as a source of micronutrients and as a neutralizer for low pH soils.

4.10. Leaching technique w.r.t pH variation

Different metal present in the fly ash has specific pH, which is able to change according to different reaction. pH has strong effect on their potential release from the fly ash. Hence, each metal had a specific pattern of leaching from ash relative to pH. In high pH (basic) solution, calcium and sodium can release easily, whereas at low pH (acidic) solution, iron and chromium released from fly ash. The release of different elements influenced by the solvent volumes, higher temperatures, longer contact times and lower solid-to-liquid ratios. Varying the condition of leaching, it is easy to concentrate Ni, Cu, Cr and Fe from fly ash. It is observed that as the pH decreases, the order of concentration of different metals is as follows [16]:

$$Ca-Na-Ni-Mg-Pb-Mn-Cu-Zn-Al-Fe-Cr$$

(most released at high pH) (most released at low pH)

Anions (Cl^-, SO_4^{2-}), oxy-anions of Se, As, Mo, B and Cr and cations (Al, Fe, Na, K, Ca, Sr, Ba, Zn, Cu, Cd and Mg) are leached from the fly ash heaps by the wastewater derived from the ash slurry which is to be collected for further concentration [17].

4.11. Froth flotation

In this technique, a mixture of fly ash and a collector was prepared and put in a mixture to ensure the uniformity. The slurry was added into the froth floatation chamber with air pump agitator. Carbon particles were coated and bound with the collector and rose to the surface after attaching with the air bubbles. **Figure 12** shows the top surface of the flotation cell and the collecting beaker when the collector was added to the fly ash sample. Carbon-rich fraction at the top was collected in a pan and the ash became residue at the bottom. Fly ash LOI reduction analysed in this process that depends on the doses of collector/frother [18].

Figure 12. (a) Froth flotation separation with collector dose and (b) bottom fraction from froth flotation.

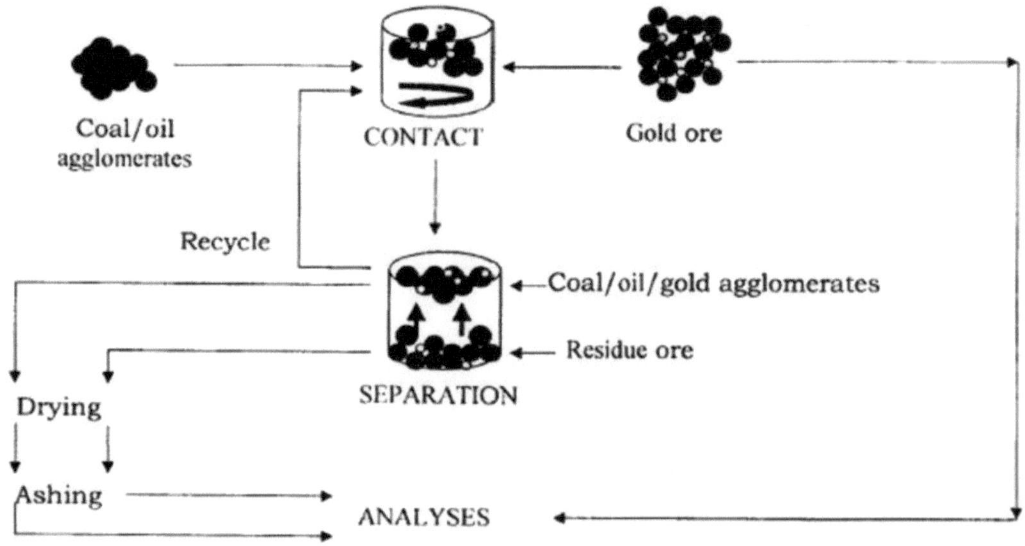

Figure 13. Flowchart of oil agglomeration process.

4.12. Oil agglomeration process

Oil agglomeration process conducted for recovery of unburned carbon. The slurry tank (**Figure 13**) is equipped with a variable speed air motor, a solvent recovery tank and a steel screen (approx. 60 mesh). In the tank, the solvent and fly ash slurry (5:1 weight ratio) was prepared. After proper agitation, the slurry was pumped into the column with air flow. The overflow unburned carbon product was collected on the screen. All the carbon recoveries were calculated on a total weight carbon basis present in the feed fly ash [19].

5. Fly ash utilization

Different technique has been developed for separation of various materials from fly ash that gives a scope for utilization of fly ash from waste materials to a valuable product. The choice of separation processes is highly influenced by associated cost and process complexity.

5.1. Treatment of acid mine drainage (AMD)

Acid mine drainage (AMD), acid/metalliferous drainage or acid rock drainage represent the outflow/downstream of acidic water from a mine site. AMD generates naturally when moisture comes into contact with pyrites and sulphidic elements (sulphide of Cu, Au, Ag, Pb, Zn) during mining results in the formation of sulphuric acid. This acidic fluid can dissolve other elements and releases toxic metals. Mostly, this acid comes primarily from oxidation of FeS_2 (iron sulphide, also called as pyrite), which is also form complex compound with various valuable metals. AMD as the coloured fluid is shown in **Figure 14**. AMD is a major problem with many ore mines where the sulphur is bound up with different metal ore. A significant number of coal mines also suffer from AMD. AMD is a major problem found in Africa, Europe, Oceania and North America. Both AMD and coal fly ash individually pose substantial

Figure 14. Liquid that drains from fly ash-filled coal mine that is highly acidic.

environmental and economic problem. In many localities (construction sites and transportation corridors), the liquid that drains from coal stocks, coal washeries, coal handling facilities and coal waste tips can be highly acidic (as shown in **Figure 1**), and this acidic fluid treated as AMD. It contains high concentrations of heavy metals such as Fe, Mn and Al and anions such as Mg^{2+}, Ca^{2+}, Mn^{2+} and $SO_4{}^{2-}$ in addition to elements such as Zn, Co, Pb, Cr and Cu, in trace concentrations, which necessitate these AMD fluids to be treated before expose to environmental use. Sulphide minerals or pyrites undergo oxidation reactions, which generate acidity and increase the concentrations of Fe, sulphate and other toxic elements in recipient water bodies (Eq. (1)).

$$\mathrm{FeS(s)} + \mathrm{O} + \mathrm{HO} \Rightarrow Fe^{+} + \mathrm{SO}^{-} + \mathrm{H} \tag{1}$$

The general reactions for this process are

$$2\mathrm{FeS_2(s)} + 7\mathrm{O_2(g)} + 2\mathrm{H_2O(l)} = 2\mathrm{Fe^{2+}(aq)} + 4\mathrm{SO_4{}^{2-}(aq)} + 4\mathrm{H^+(aq)} \tag{2}$$

$$2\mathrm{Fe_2S(s)} + 7.5\mathrm{O_2(g)} + 7\mathrm{H_2O(l)} \Rightarrow 2\mathrm{Fe(OH)_3} + 4\mathrm{H_2SO_4} \tag{3}$$

$$4\mathrm{FeS_2} + 15\mathrm{O_2} + 14\mathrm{H_2O} \Rightarrow 4\mathrm{Fe(OH)_3} + 8\mathrm{H_2SO_4} \tag{4}$$

Oxidation from sulphide to sulphate solubilizes the ferrous iron, which is subsequently oxidized to ferric iron:

$$4\mathrm{Fe^{2+}(aq)} + \mathrm{O_2(g)} + 4\mathrm{H^+(aq)} = 4\mathrm{Fe^{3+}(aq)} + 2\mathrm{H_2O(l)} \tag{5}$$

Again, the ferric cations produced can also oxidize additional pyrite and reduce into ferrous ions either spontaneously or catalysed by microorganisms:

$$\mathrm{FeS_2(s)} + 14\mathrm{Fe^{3+}(aq)} + 8\mathrm{H_2O(l)} = 15\mathrm{Fe^{2+}(aq)} + 2\mathrm{SO_4{}^{2-}(aq)} + 16\mathrm{H^+(aq)} \tag{6}$$

Some bacteria act as a catalyst in generating ferric hydroxide precipitates known as 'yellow boy', which is also a contributor for the formation of AMD.

$$\mathrm{Fe^{3+}} + 3\mathrm{H_2O} \rightarrow \mathrm{Fe(OH)_3} + 3\mathrm{H^+} \tag{7}$$

All the above reactions result the release of H^+, which decrease the pH value that maintains the solubility of the ferric ion.

The water's acidity, toxicity and the yellow coating of iron oxide affect water streams, rivers and aquatic life mainly due to AMD. It is very important task to manage the AMD polluted mine water in expense of active and passive remediation engineering technologies. It is found that AMD can be neutralized by introducing the fly ash. AMD was reacted with coal ash in a two and a half months' equilibration time to produce neutral and alkaline process waters and remove major and minor elements from solution. Concentration of various element depends on their solubility control exist in the solution towards neutralization. Generally, $SO_4{}^{2-}$ concentrations are not affected by neutralization unless the formation of mineral saturation due to gypsum [20]. Hence, SO_4^{2-} plays an overall indicator for acid generation after neutralization. Sulphate concentrations were found to be controlled by the precipitation of gypsum ($CaSO_{4.2}$ H_2O) at low pH, ettringite ($Ca_6A_{12}O_6$ $(SO_4)_{3.32}H_2O$) at high pH. Hydrolysis reaction among Fe^{3+}, Al^{3+}, Fe^{2+} and Mn^{2+} decreases the pH and dissolve the oxide constituents. Toxins in AMD, including radionuclides, high sulphate levels and numerous other toxic elements (such as Al, Fe, Mn, Hg, As), result a highly acidic fluid. The soluble bases (oxides) in coal fly ash and hydrolysable constituents in AMD results about the final solution weather a dominant acid or basic character at a given contact time.

5.2. Cement

Fly ash has pozzolanic properties and can replace 30% by mass of the Portland cement that utilized for manufacture of the cement [21]. Cement industries utilize more than 50% of fly ash that can add strength and durability to the concrete. It is expected to utilize more than 75% by 2020 with increasing the demand of cement in infrastructure industries. Chemical reactions involve in cement are as follows:

$$2(3CaO \cdot SiO_2) + 6H_2O \rightarrow 3CaO \cdot 2SiO_2 \cdot 3H_2O + 3Ca(OH)_2 \quad (8)$$

$$2(2CaO.SiO_2) + 4H_2O \rightarrow 3CaO \cdot 2SiO_2 \cdot 3H_2O + Ca(OH)_2 \quad (9)$$

In cement and fly ash matrix, lime is released during hydration of tricalcium silicate ($3CaO{\cdot}SiO_2$) and dicalcium silicate ($2CaO{\cdot}SiO_2$). The tricalcium silicate is more suitable for stabilization with fly ash.

5.3. Waste solidification and stabilization

Stabilization is a process that utilizes additives to reduce the hazardous nature of a waste in the environment. The fly ash can play a major role in the stabilization/solidification process [22]. Fly ash mixed with cement can be used to solidify and stabilize heavy metal sludge. An optimum mixture composed of 45% fly ash, 5% cement and 50% of the industrial sludge could provide the required solidification and stabilization. Fly ash and sludge mixture (without cement) possess the necessary compressive strength for landfill. Generally, there are two best results of mixture used for the production of brick: (i) pulverized sludge + brick clay + fly ash and (ii) pulverized sludge + fly ash + lime.

5.4. Mine void filling

Embankment fill or backfill typically use to create a strong compact base that has low modulus of elasticity (≈50 kg/cm^2), better permeability (10^{-4} to 10^{-6} cm/s), higher shear strength, higher compressibility and less stiff than the magnitude of the surrounding rock. The purpose of the backfill is to restrict the rock stresses and improve the load shedding to mine pillars that leads to less deterioration for ground conditions. Local soil is too weak for backfill and is replaced by compacted fly ash materials. Fly ash embankment provides a better bearing capacity and strength [23]. Generally, fly ash placed and compacted with moisture content to reduce construction time and equipment costs with increasing compact strength. Fly ash embankment has a loose dry density as low as 650–810 kg/m^3 (40–50 lb/ft^3). But when it compacted with optimum moisture content (≈20–35%), the dry unit weight of fly ash may be greater than 1620 kg/m^3 (100 lb/ft^3). **Figure 15** shows the mine void filling, in which minimum 0.5 m general soil is covered in between the fly ash layer (1–3 m) for better strength.

5.5. Structural fill and cover material

It is very logical and environmental friendly to utilize and dispose of large volume of fly ash in the form of structural fill or cover material. Fly ash can be utilized to fill in the internal-void of wall, in structural support, bridge abutments. The density, particle size and strength are the more important properties before using as a cover material. Generally, fly ash compacted to a density range (1120–1520 kg/m^3) for cover material.

5.6. Concrete

Fly ash has cementitious properties that used for production of Portland cement concrete. Note that 30–50% masses of fly ash have been used in massive structures such as foundations and dams and 40–60% of fly ash can be used in structural applications, producing concrete with good mechanical properties and durability. A well-proportioned fly ash concrete mixture will give improved workability, cohesiveness and reduces segregation when compared with a Portland cement concrete of the same slump. Fly ash concrete flows and consolidates are more efficient than a conventional Portland cement concrete when vibrated. Due to spherical particle shape of fly ash, it is very easy to pump and reduce wear on equipment [24].

5.7. Floor blocks and tiles

Floor blocks are the aerated autoclaved concrete made from mixture of fly ash, cement, lime, gypsum and alumina. These blocks are steam cured to get the desire strength, colour, etc. These blocks have better properties such as non-toxicity, earthquake resistant, noise insulation, lightweight, thermal insulation, high compressive strength, high strength to weight ratio, lower water absorption and dimensional accuracy. These blocks can be used in airfield, sub-base and base requirements of all classes of road, side walls, commercial pavements, gardens, parking areas, drive ways, etc. (**Figure 16**). Fly ash is also used for manufacturing superior finish and high strength reflective tiles can be made in different shapes and thicknesses [25].

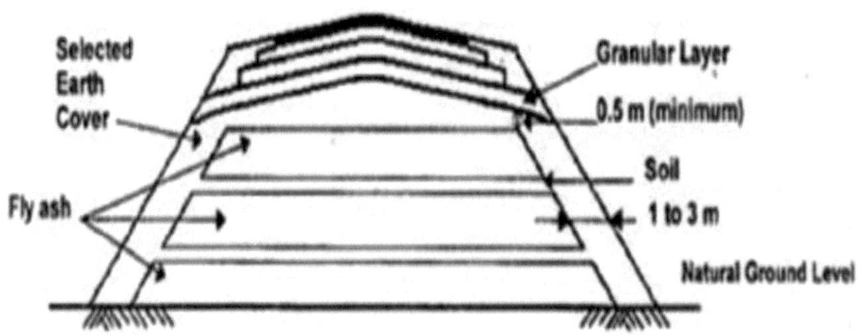

Figure 15. Cross-section of void fills with alternate fly ash and soil.

5.8. Road making

Many gravel roads suffer from reduced bearing capacity mainly during adverse climate conditions (higher temperature or rain). Thus, the soil frost periods will become shorter that lead to degradation of gravel roads. To avoid the reduced bearing capacity, the gravel roads can be stabilised by using fly ash and can give high compression strength. Fly ash can be compacted using moisture content, which is easy to compact using both vibratory and static rollers, and no large lumps to be broken down. Pozzolanic hardening characteristic gives additional strength to the road pavements. Fly ash decreases swelling potential of expansive soils. **Figure 17** shows different fly ash roads which have good surface finish and strength.

5.9. Land development in agriculture

The successful application of fly ash has done in the agriculture field. Fly ash acts as a soil modifier and source of micronutrient. Various seasonal crops have been grown and harvested and there is a significant increase in yield of different crop. It is evident from the crop yield that the addition of 10–200 tonnes fly ash per hectare increased the yield from 5 to 40%. Moreover, fly ash of 10–25% level could be used as a good potting mix material in forestry nurseries to produce hardy seedlings. Fly ash input in agriculture land has proved to minimise the addition of chemical fertilizers and plant growth promoters. But the limiting concentration of fly ash in cultivating land is 25% and used as a secondary nutrient without causing any harm effect on soil health. **Figure 18** shows the cultivation improvement by using the fly ash as a secondary nutrient.

Figure 16. Fly ash floor bricks and tiles.

Figure 17. Utilization of fly ash in different road.

5.10. Reclamation of sea

Many Indian cities are facing shortage of land, and land reclamation from sea can give great success for these cities [26]. A large quantity of fly ash permitted to reclaim the land from sea, basically in coastal urban cities. In 2003, Govt. of India has permitted the use of fly ash for reclamation of sea. Subject to the rules made under the Environment Protection Act (1986), reclamation of sea shall be a permissible method of utilization of fly ash. Land reclamation includes maintaining water and air quality, erosion and damage to land properties, minimizing

Figure 18. Agricultural development by fly ash in Odisha.

Figure 19. Residential complex in Palm Jumeirah, Dubai.

flooding, wildlife and aquatic habitats caused by surface mining. Singapore and Dubai have conducted this type of sea reclamation in a large scale (**Figure 19**). The three alternative ways have been investigated to utilize the waste in sea reclamation projects: (1) Singapore and Japan-based technology, (2) plasma gasification-based technology and (3) strengthened sediment-based land reclamation technology.

6. Conclusion

This chapter discusses the idea about recovery of different elements from fly ash and utilization of fly ash. The approach taken in this chapter has interesting implications for different technology for fly ash beneficiation and fly ash utilization. Lime, sand, aggregate, metals and rare earths recover from fly ash can fulfil demand for natural resources. Fly ash constituents depend upon the source industries specific coal bed but may include substantial amounts of oxide elements and heavy metal. Due to the treatment of fly ash by different extraction techniques, fly ash became environmental friendly. Fly ash utilization as given in this chapter is very essential for future study. Most important thing is the efficient treatment of AMD. Moreover, selective sequential extractions by fly ash help to recover different value-added materials from AMD with their neutralization reactions.

Author details

Ajit Behera[1]* and Soumya Sanjeeb Mohapatra[2]

*Address all correspondence to: ajit.behera88@gmail.com

1 Department of Metallurgical & Materials Engineering, National Institute of Technology, Rourkela, Odisha, India

2 Department of Chemical Engineering, National Institute of Technology, Rourkela, Odisha, India

References

[1] Mishra BK, Das B, Biswal SK, Reddy PSR. Overview of beneficiation, utilization and environmental issues in relation to coal processing. Proceedings of the Indian National Science Academy. 2015;**81**(4):725-737. DOI: 10.16943/ptinsa/2015/v81i4/48293

[2] Malik A, Thapliyal A. Eco-friendly fly ash utilization: Potential for land application. Journal Critical Reviews in Environmental Science and Technology. 2009;**39**:333-366. DOI:10.1080/10643380701413690

[3] Bhattacharjee U, Kandpal TC. Potential of fly ash utilisation in India. Energy. 2002;**27**: 151-166

[4] Dhadse S, Kumari P, Bhagia LJ. Fly ash characterization, utilization and Government initiatives in India Œ A review. Journal of Scientific & Industrial Research. 2008;**67**:11-18

[5] Vassilev SV, Vassileva CG. A new approach for the classification of coal fly ashes based on their origin, composition, properties, and behaviour. Fuel. 2007;**86**:1490-1512

[6] Wecharatana M, Liskowitz JW. Properties of high-performance fly ash concrete. The American Concrete Institute. 1996;**159**:429-444. https://www.concrete.org/publications/internationalconcreteabstractsportal.aspx?m¼details&ID¼16769

[7] van Riessen A, Chen-Tan N. Beneficiation of collie fly ash for synthesis of geopolymer: Part 1-beneficiation. Fuel. 2013;**106**:569-575

[8] Kiattikomol K, Jaturapitakkul C, Songpiriyakij S, Chutubtim S. A study of ground coarse fly ashes with different finenesses from various sources as pozzolanic materials. Cement and Concrete Composites. 2001;**23**(4–5):335-343

[9] Hemmings R T, Berry E E. Speciation in size and density fractionated fly ash. MRS online proceedings library. 1985;**65**:91. DOI: https://doi.org/10.1557/PROC-65-91

[10] Shoumkova AS. Magnetic separation of coal fly ash from Bulgarian power plants. Waste Management & Research. 2011;**29**(10):1078-1089

[11] Kruger RA. Fly ash beneficiation in South Africa: Creating new opportunities in the market-place. Fuel. 1997;**76**(8):777-779

[12] Soong Y, Schoffstall MR, Link TA. Triboelectrostatic beneficiation of fly ash. Fuel. 2001;**80**(6): 879-884

[13] Emre Altun N, Xiaob C, Hwangb J-Y. Separation of unburned carbon from fly ash using a concurrent flotation column. Fuel Processing Technology. 2009;**90**(12):1464-1470

[14] Gray ML, Champagne KJ, Soong Y, Killmeyer RP, Maroto-Valer MM, Andrésen JM, Ciocco MV, Zandhuis PH. Physical cleaning of high carbon fly ash. Fuel Processing Technology. 2002;**76**(1):11-21

[15] Canty GA, Atalay A, Laguros JG, Robertson J, Pandey KK. A preliminary assessment of utilizing fluidized bed ash in landfill liner applications, Journal of Environmental Science and Health. Part A: Environmental Science and Engineering and Toxicology, 1995;**30**: 439-459

[16] Karapanagioti HK, Atalay AS. Laboratory evaluation of ash materials as acid-disturbed land amendments. GlobalNEST International Journal. 2001;**3**(1):11-21

[17] Gitari WM, Petrik LF, Key DL, Okujeni C. Partitioning of major and trace inorganic contaminants in fly ash acid mine drainage derived solid residues. International Journal of Environmental Science and Technology. 2010;**7**(3):519-534

[18] Demirbas A. Demineralization and desulfurization of coals via column froth flotation and different methods. Energy Conversion and Management. 2002;**43**(7):885-895

[19] Graya ML, Champagnea KJ, Soonga Y, Finseth DH. Parametric study of the column oil agglomeration of fly ash. Fuel. 2001;**80**(6):867-871

[20] Geldenhuis S, Bell FG. Acid mine drainage at a coal mine in the eastern Transvaal, South Africa. Environmental Geology. 1998;**34**(2):234-242

[21] Canpolat F, Yılmaz K, Kösec MM, Sümer M, Yurdusev MA. Use of zeolite, coal bottom ash and fly ash as replacement materials in cement production. Cement and Concrete Research. 2004;**34**(5):731-735

[22] Dermatas D, Meng X. Utilization of fly ash for stabilization/solidification of heavy metal contaminated soils. Engineering Geology. 2003;**70**(3–4):377-394

[23] Martin JP, Collins RA, Browning JS, Biehl FJ. Properties and use of fly ashes for embankments. Journal of Energy Engineering. 1990;**116**(2). DOI: http://dx.doi.org/10.1061/(ASCE)0733–9402(1990)116:2(71)—see more at: http://ascelibrary.org/doi/pdf/10.1061/(ASCE)0733-9402(1990)116:2(71)#sthash.PYUoT92P.dpuf

[24] Al-Amoudi O, Maslehuddin M, Asi I. Performance and correlation of the properties of fly ash cement concrete. Cement, Concrete and Aggregates. 1996;**18**(2):71-77. https://doi.org/10.1520/CCA10153J. ISSN 0149-6123

[25] Blissett RS, Rowson NA. A review of the multi-component utilisation of coal fly ash. Fuel. 2012;**97**:1-23

[26] Michalkova E, Schwarz M, Pulišová P, Máša B, Sudovský P. Metals recovery from acid mine drainage and possibilities for their utilization. Polish Journal of Environmental Studies. 2013;**22**(4):1111-1118

Chapter 8

Thickener Water Neutralization by Mid-Bottom and Fly Ash of Thermal Power Plants and CO_2: Organic Humate Mud of AMD Treatment for Remediation of Agricultural Fields

Yıldırım İsmail Tosun

Additional information is available at the end of the chapter

http://dx.doi.org/10.5772/intechopen.69927

Abstract

Coal mining sites and copper sulphide mines caused polluting acidic mine waters. Acidic mine drainage (AMD) into the natural streams and agricultural field irrigation waters causes environmental pollution with dissolved contents having heavy metals or acidic aqueous solutions of habitats and living fish in the local ponds and irrigation ponds. It is necessary to collect the polluted mine waters in the impermeable ponds and treat or neutralize the acidic character by the aeration treatment and subsequently discharge to the stream waters. This study introduced chemical neutralization and aeration treatment of polluted acidic waters in Şırnak Coal Mine site and Siirt Copper Concentrator site as happens in wastewaters of chemical industries and textile processing plants. The construction engineering of collector ponds was also so much critical in the point of cost of a case remediation model and muddy fertilization by Şırnak Thermal Power Plant waste, bottom ash, and fly ash in agricultural fields soil in Şırnak Coal Mine site, and how these facts may affect the possibility of remediation in the future were discussed. Organic fertilizer matter of fumic acid and humic acid containing humate formed. Soil remediation process for both muddy bottom precipitates of treatment pools and aeration pools was collected and treated by CO_2 saturatedwaters. The comparison of the common methods is described and discussed with fly ash utilization and CO_2 gas of power plants containing alkali matter improved neutralization and carbon matter of soils.

Keywords: acid mine drainage, geochemistry, soil remediation, restoration, construction

1. Introduction

Acid mine drainage brings out great environmental issue of acidic metallic sulfate rich wastewaters. Those type wastewaters are also the wastewaters of a variety of industrial operations such as galvanizing, leaching, and the scrubbing of flue gases at thermal power plants [1–5]. The fundamental reasons of such wastewaters are under the agricultural and environmental concern occurring in the mining industry. Waters draining active and abandoned mines and mine wastes are often below pH 4 as low as sulphatic polluted (extremely so acidic). Such waters typically cause an additional risk to the environment by the fact that they often contain elevated concentrations of metals (iron, aluminium and manganese, zinc and copper, possibly Hg and Pb, and other heavy metals) and metalloids (arsenic and sulfur). The environmental issue caused by mine water discharges is difficult to manage accurately. It is generally preferable to prevent acidic water source by collector pools and settling the mud, although not always pragmatic, to preclude the formation of acidic mine drainage (AMD) in the first instance. Such techniques are known collectively as bsource control the precautive measures (**Figure 1**) and will be described only briefly [6–9].

In the formation of AMD, both oxygen and water affect to metal dissolution. The proposed precautions follow that by excluding either of these, it is necessary to prevent or minimize AMD production. The common method is practiced by sealing hazardous sites. The oxygen available in the slurry is utilized by oxidizing bacteria present, and completion of dissolved O_2 by mass transfer, and diffusion is delayed by sealing the wastewaters. This is seen in the location of all contaminated seepage waters. Water storage was used for disposing and storing mine tailings that are potentially acid producing [10]. However, the main goal is to reduce the contact between the minerals and dissolved oxygen. Wastewater is chemically covered, or their reactivity is decreased by covering the waste material with a layer of sediment clay or organic material, beneficial on oxygen elimination and improved carbon dioxide gas diffuse [11–13].

The additives such as clay bentonite or diatomite provide protective cover on the tailings because of the actions of slurry mixing (**Figure 2**). Clay top covers of reactive wastes and sulphidic mineral mixtures may cause clayish organic layer. The sealing layer that covers the waste slime is usually constructed from clay. In wet and dry conditions, drying and cracking of the cover might present less effective precaution at the urbanization zones [14, 15].

Figure 1. The lake area of AMD in Çan and satellite image AMD lake in the Sırnak Coal Mine Site, pool remediation [9].

http://dx.doi.org/10.5772/intechopen.69927

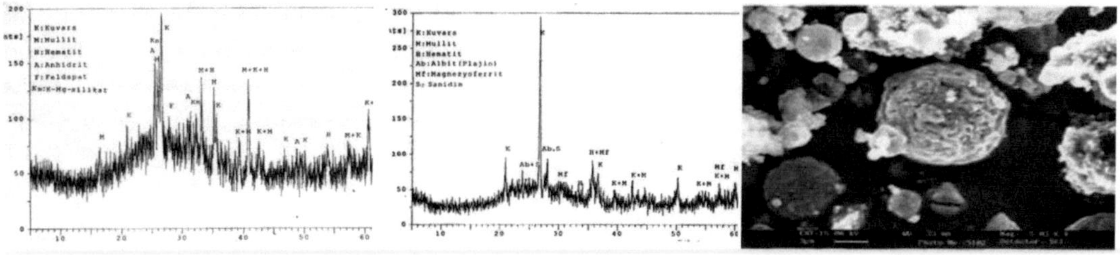

Figure 2. The X-ray diffractometer diagrams of Orhaneli and Seyitömer Fly Ash, SEM picture of fly ash in the Şırnak Coal Mine Site [6, 9].

The unsettling rust color of the water also served to make acid mine drainage an easy/prime target for environmental agencies around the world. However, when mines are closed and abandoned, and pumps are turned off, the rebound of the water table can lead to contaminated groundwater being discharged, sometimes in a catastrophic event such as the one that happened at the Wheal Jane mine in 1992 when a range of contaminants entered the environment [1, 5–14, 16]. As the water that refills the mine dissolves any acidic salts that have built up on the pore spaces of the exposed walls and ceilings of underground chambers.

Another suggested approach for minimizing AMD production is to blend acid-generating and acid-consuming materials, producing environmentally benign composites [16, 17]. A variant on this case is to add solid sorbents such as phosphates salts, fly ash, and apatite to acidic mine waste in order to precipitate iron (III) as ferric phosphate, thereby reducing its potential to act as an oxidant of sulfide minerals. The magnesium or calcium salts, phosphate minerals, or availability of soluble phosphate (together with hydrogen peroxide) eliminates oxidation of pyrite, producing ferric iron, which reacts with the phosphate to produce a surface protective coating of ferric ion [18]. An alternative technique involving the formation of an iron oxide/silica coating on pyrite surfaces has also been described. The kinetics of microorganism growth and act was critical in a month time period and over. However, certain time of season period of temperature and oxidation time was observed as followed 6 months' time duration by moisture even in the drought zone.

The lithotrophic iron- and sulfur-oxidizing bacteria had a main role in generating acidic waters in various laboratory tests and geothermal springs. The inhibitors or conditions eliminated their activities in mineral slimes and tailings. The anionic surfactants such as sodium dodecyl sulfate (SDS) are highly toxic to this group of bacteria [12, 13]. Chemolithotrophic thiobacillus type organisms needed low pH under 2 and higher temperatures for thermophilics for sulfuric acid production in the commence of oxidation reactions on complex sulphide ores.

However, the effectiveness of biocide applications has been found to be highly variable affording, at best, only short-term control of the problem and requiring repeated applications of the chemicals [19]. While the practical difficulties prevailed in inhibiting the formation of AMD at source, often, the method is to minimize the impact that this polluting water has on receiving streams and rivers, and the wider environment; such an approach involves. It has been divided into active and passive processes. In general, concerned main effect of alkali materials is to neutralize acidic mine waters and precipitation of metal salts, by-product mud to the use of natural reconstruction of ecology in this studied project. The passive irrigation channels and pools were feasible for high flow rate slurries in the site requiring relatively little maintenance than active systems. In truth, the practiced passive treatment method in the projected site required a certain amount of maintenance costs.

2. Utility of fly ash and mid-bottom ash

Fly ash is formed by keeping the particles in the flue gases in the electric filters while burning the pulverized coal in the thermal power plant boilers alone. It is possible to obtain the following three different yields from the coal. Base Ash, which is relatively coarse and cannot be transported by flue gas. Mid Ash, which burned in cyclone type boilers. Fly Ash, which is very fine grained and transported with flue gases. The fly ashes are collected mechanically and electrostatically and stored in the vicinity of the plant or in other convenient places. For the purpose of collecting ash, the ashes accumulate in large areas and become a problem for the plant management.

The amount of fly ash yielded in the world is about 600 million tons per year. Almost 26 thermal power plants are currently operating in Turkey, including Afşin-Elbistan, Çatalağzı, Kangal, Kemerköy, Orhaneli, Seyitömer, Soma, Tunçbilek, Yatağan, and Yeniköy. The average annual fly ash production from these plants is about 13 million tons, but the annual rate varies with the introduction of natural gas power plants. One way of reducing the country's external dependence on energy production is to use low-calorie lignite coals in thermal power plants, which cannot be used in other parts of the industry. Therefore, the annual fly ash amounts are expected to increase further in the future. The X-ray diffractogram of the mineralogical composition of the Orhaneli Thermal Power Plant fly ash is given below. According to this, it is understood that the main phase of the Orhaneli fly ash is the glassy phase. In addition to the glassy phase, it is also found that there are crystallized phases such as quartz, mullite, hematite, anhydrite, sanidine, and K-Fe silicate. The main phase is the glassy phase in the X-ray diffractogram of the Orhanli fly ash, the background is generally in a high position. However, especially, in the 23–27°2θ, the background showed a significant increase. In the chemical analysis, as the SiO_2 48.53% and Al_2O_3 24.61% were found in the volumetric cluster, the shift toward 27°2θ was more Al_2O_3 and CaO (9.48%) (**Table 1**). The X-ray diffractogram of the mineralogical composition of the Seyitömer Thermal Power Plant fly ash is given in **Figure 2**. As can be seen in the volcanic core, there are primarily glassy phases, quartz, mullite, magnesioferrite, hematite crystals, and alkali feldspars as minor. In the X-ray diffractogram of the Seyitömer fly ash, the glassy phase gave the maximum between 22 and 25, and the view

	Afsin Elbistan	Orhaneli	Yatağan	Seyitömer	Sırnak	ASTM C616 F	ASTM C616 C
SiO_2	34.11	48.71	52.3	54.11	40.71		
Al_2O_3	10.58	24.53	23.2	20.58	11.53		
Fe_2O_3	4.33	7.62	5.9	9.33	5.62		
S + A + F	74.52	81.46	81.7	84.52	88.6	≥70	≥50
CaO	24.72	9.56	10.1	4.72	19.56		
MgO	3.33	2.41	2.4	4.33	2.41	≤5	≤5
SO_4	2.72	3.02	1.5	0.72	1.02	≤5	≤5
KO	1.10	2.44	2.8	2.10	2.44		
Na_2O	0.57	0.37	1.9	0.67	0.55		
Ignition Loss	4.22	1.74	0.9	3.22	4.74	≤ 6 ≤ 10(TSE639)	≤ 6 ≤ 10(TSE639)
Cl	0.006	0.005	0.0	0.006	0.0		
Free CaO	1.72	1.13	0.3	1.72	3.13		
React. SiO_2	19.8	34.12	30.1	39.8	34.12		
React. CaO	13.16	6.72	7.1	2.16	12.72		

Table 1. The chemical analysis of utilized fly ash in the clarification of water pool in the Sırnak Coal Mine Site, fishery pool remediation [6, 9].

was glassy rich. In this case, the vitreous phase shows a siliceous character due to the quartz crystal being close to the maximum peak (**Figure 2**) [6, 9].

As a result of the morphological examination of the Orhanli flywheel, it was observed that the majority of spheroids ranging in size from 0.5 to 15 μ were regular spherical particles. Full spherical vitreous solid particles come into play due to the rapid cooling of the droplets that form as a result of the majority (clay) of the minerals in the coal due to the high temperature reached during the burning of the car. In addition to these, plospores with a hollow cavity and a cluster of microspheres are also encountered. In addition, conchoidal quartz crystals, surface anhydrite-coated particles, and fine prismatic mullite crystals have been identified in the microstructure. The result of the separation of the reactive glassy phase after the dissolution of the mullitic crystals in the ash acid has been revealed. The finely dispersed mullite is formed by partial crystallization of the vitreous phase consisting of aluminosilicate; in the senosphere, Quartz crystals have been added to the volatile compounds during their burning due to their high thermal stability, leaving the environment and partially protecting the original structures (conkoidal). Some spheres have accumulated anhydrous (cubic) and hematite (fine crystallized) deposits on the surface. The chemical characteristics of Şırnak and other fly ash and mid-bottom ash are illustrated in **Figure 3** used in AMD remediation. The increased iron and zinc contents of fly ash for neutralization decreased the effectivity in waste treatment due to acidic manner of composition dissolved in wastewater slurries by diminishing alkali reactions in the pool, while lighter heavy metals such as Pb and Zn should preferred in neutralization act on wastewaters in Şırnak Coal Mine site [6, 9].

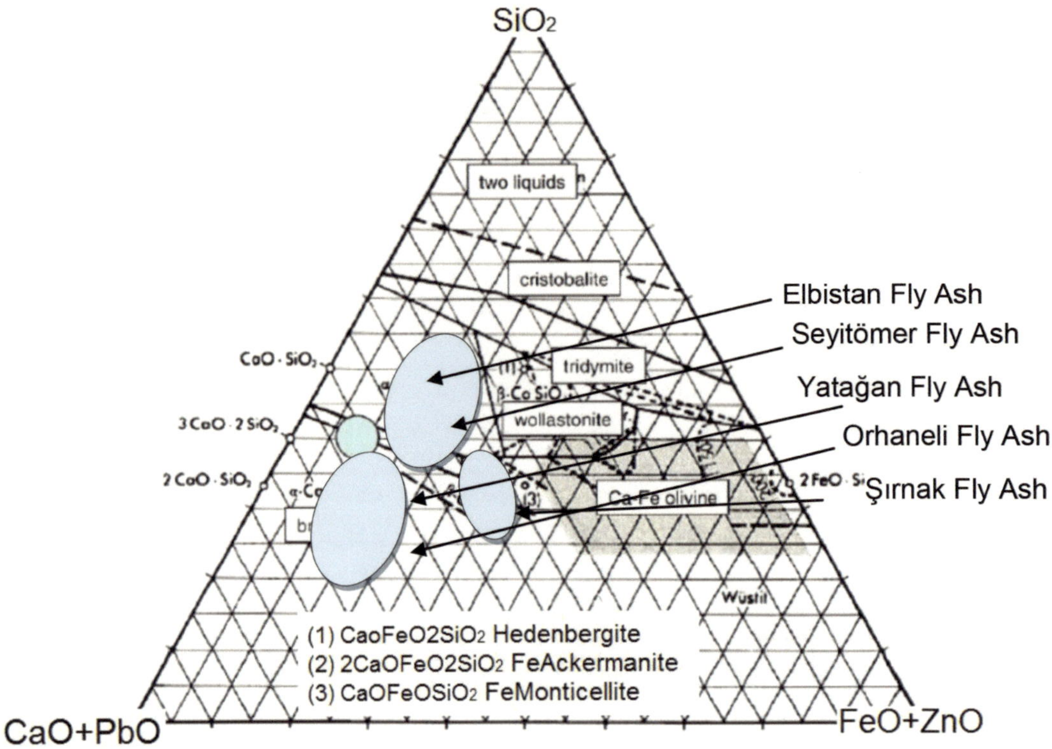

Figure 3. The chemical phase diagram of Sırnak Coal Mid Ash and Fly Ash used in remediation.

3. AMD waste treatment

The chemical reactions are following the paths as seen in **Figure 4**. The Fe and Cu sulfur matter oxidation potential for the first are shown, respectively. Many aquifers have oxidation states low enough to reach the threshold for these reactions to occur.

The value of pH required to induce precipitation may increase by neutralization in aguous euphotropic waters and lake. However, despite the general importance of pH in an AMD treatment systems, it is not the only fundamental neutralization for metal removal, but also ionic strength, temperature, Eh, and concentrations of suitable complexing agents may change the precipitation. Treatment sludges therefore include the well-known crystalline form of goethite but consist mainly of highly amorphous oxy and hydroxyl iron oxides.

3.1. Active technology

The common method used to treat acidic effluents is an active treatment process involving neutralization work reagent [20–25]. Addition of an alkali to AMD wastewaters easily changes the pH and commence of chemical oxidation on active aeration. While addition of a chemical prompter such as hydrogen peroxide is also beneficial, it causes the common metals present in solution to precipitate as hydroxides and carbonates. The active neutralization by alkali

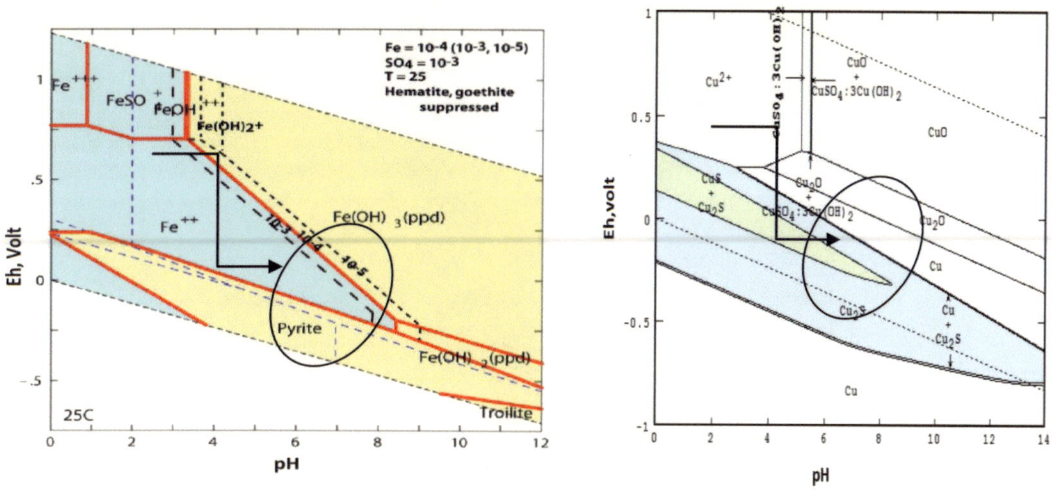

Figure 4. The Eh-pH phase diagrams in the Fe-S-O-H and Cu-S-O-H system with lg [Fe], lg [S] and Eh as coordinates at the pH 5 and pH 7.5 [19].

matters is governed by the reactions given below Eqs. (1) and (2), as respected oxidation ferrous iron to ferric state was rate controlled on the concentration in the solute. The direct neutralization could be managed by settling of wastewaters during clarification by alkali sorbents (**Figures 5** and **6**). The lamelle settlers provide much settling area per flow rate wastewaters containing less solids.

$$4Fe^{2+} + O_2 + 4H_2 \rightarrow 4Fe^{3+} + 2H_2O \quad (1)$$

$$4Fe^{3+}12H_2O \rightarrow 4Fe(OH)_3 + 12H_2 \quad (2)$$

3.2. Passive technology

Ammonification occurs as an alkali yield process. Due to the lack of materials such as nitrate, the nitrification and ammonification are so slow and have minor importance in AMD waters and environments [26–32]. However, since both ferric iron and sulfate tend to be highly

Figure 5. The thickener clarification of AMD water in the Sırnak Coal Mine Site for fishery pool remediation e^-.

Figure 6. The Lamelle Thickening in AMD Water Treatment.

abundant in AMD seepage, alkali resulting from the reduction of these two species, a weak base (bicarbonate) and producing a strong base (hydroxyl ions), also generate net alkalinity (Eq. 3). The indirect acid production was relatively high at higher pH levels over 5 with dissolution of heavy metals in sulphide minerals, and neutralization by alkali matters governs the dissolution by the reactions given below:

$$SO_4^{=} + 2CH_2O + 2H^+ \rightarrow H_2S + 2CO_2 + 2H_2O \quad (3)$$

The sulfide produced is strongly reactive toward heavy metals as given in Eqs. (4) and (5):

$$Fe^{2+} + H_2S \rightarrow FeS + 2H^+, \quad (4)$$

$$Zn^{2+} + H_2S \rightarrow ZnS + 2H^+, \quad (5)$$

which form very insoluble sulfide compounds. FeS is unstable relative to pyrite and the further reaction, which is an oxidation of S^{2-} to S^-, as given in Eq. (6):

$$FeS + S \rightarrow FeS_2 \quad (6)$$

generated in the late muds close to the settled mud–water interface. ZnS and PbS, in the sulphide complex structure, are much stable and retain S in the −2 state. However, the sulfate part of reaction (Eqs. (7) and (8)) may cause redox effect an oxidation. Then

$$H_2S + 4H_2O \rightarrow SO_4^{2-} + 10H^+ + 8e^-. \quad (7)$$

$$\text{Aqeous } 6HCO_3^- \rightarrow 6H_2O + C_6H_{12}O_6 + 6O_2 + 6OH^- \quad (8)$$

The precipitation of ferric iron does not decrease solution acidity, and the reverted ferrous iron as indicated in Eq. (9) generally is supplied by inorganic surface.

$$Fe(OH)_3 + 3H_2 + e^- \rightarrow Fe^{2+} + 3H_2O \quad (9)$$

Bacteria that catalyze the dissimilatory reduction of sulfate to sulfide generate alkalinity by transforming a strong acid (sulfuric) into a relatively weak acid (hydrogen sulfide; Eq. (10)).

$$(SO_4)^{2-} + 2CH_2O + 2H_2 \rightarrow H_2S + 2H_2CO_3 \quad (10)$$

Besides the neutralization of AMD, the passive act results an increase in pH, the reduction of sulfate is an important mechanism in eliminating toxic metals from AMD seepage, since as many (e.g., lead, zinc, copper, and cadmium) form highly insoluble sulfides (e.g., Eq. (11)).

$$Zn^2 + H_2S \rightarrow ZnS + 2H_2 \quad (11)$$

Biological oxidation of ferrous iron to ferric (which is highly insoluble above pH 2.5) is the other major metal-immobilizing process that occurs in aerobic wetlands and bioreactors. As shown in **Figure 7**, most microorganisms' act is a slow passive system. The constructed wetlands or compost bioreactors have been unfeasible for high flow rated wastes in full-scale. The basic advantage of passive bioremediation is relatively low maintenance costs. Due to the fact that the mud products of sludge are retained within the wetland sediments, the treatment handling is expensive. The landfill area available to install is higher than chemical treatment systems, and the long-term accumulation of wastes within the pools is uncertain [2, 33–37].

Passive bioremediation systems that utilize a combination of aerobic and anaerobic wetlands have been used for full-scale treatment of AMD. An example is the acid reduction using microbiology (ARUM) system [7, 9, 38–41]. Bioremediation and anoxic treatment of Şırnak Coal Mine acidic waters were carried out by limestone, fly ash, and waste coal layer aeration pools as illustrated in **Figure 7** [7, 9]. While there is insufficient alkalinity in the mine water to prevent a significant fall in pH, the kinetics of oxidation is controlled by the Eq. (12):

$$\frac{dC(Fe_2O_3)}{dt} = k^{1/2}CO_2C(Fe)^{2+}C(SO_4)^{2-} \quad (12)$$

This iron is oxidized and precipitated and followed by AMD passes first through ARUM cells, alkali and sulfide are generated. The ARUM systems showed to be effective in treating AMD in high latitude and subtropical locations [7, 42]. The passive treatment plant at the Wheal Jane site is also a composite system. The by-product of AMD treatment, zinc sulfide produced fed into the pool and reduced into the metallic zinc product from the plant [36, 43–45].

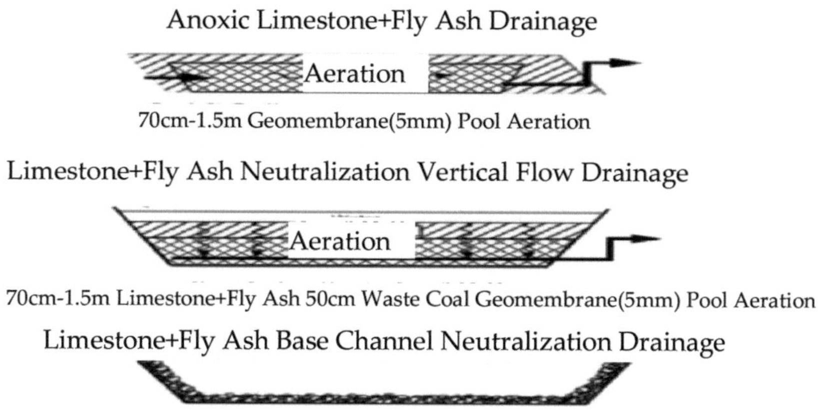

Figure 7. The Methods of Passive Limestone and Fly ash Use in AMD water Neutralization.

Sulfate-reducing bacteria (SRB) are heterotrophic bacteria and, the iron-oxidizing acidophiles, require provision of organic material such as carbon and sulfur. Hydrogen in the form of ethanol may substitute as energy for sulfate reduction and reductive act in acidic mine drainage waters as given in Eq. (13).

$$SO^{2-} + 4H_2 \rightarrow H_2S + 4H_2O \tag{13}$$

The use of hydrogen is advantageous instead of high sulfate loadings and results in lesser bacterial organic biomass contamination. However, hydrogen use in natural full scale, at high flow rated wastewaters, may not be feasible. In both cases, carbon dioxide is also produced, and some SRB can fix this as their source of carbon [46–50].

4. Projected case for Şırnak coal site and Siirt copper mine AMD treatment: Organic soil mud

This investigation of water clarification, neutralization, and organic humate mud yield was extremely useful in wastewater management strategies by reclaiming existing agricultural irrigation and water logged areas. The problems of water use and quality mostly exist in the irrigated areas like in South Eastern rocky plains of Şırnak, Turkey. The climate change and groundwater changes generally results in over irrigation, seepage losses through channel and distributions, poor water management practices, and inadequate control of drainage system. Analysis of high water table in water logged areas and drainage of irrigated areas have not been paid adequate attention in the planning and management of water resources, partly due to lack of requisite data and partly due to flood and contamination in the country. In order to develop suitable water management strategies, the area needed to control the extent of contamination and waste logging in the area. Geographic information system (GIS) may facilitate the reconstruction of the ecological environment but also to accommodate the sustainable development of the water resources and wastewater. For the treatment of wastewaters seepages, leaks of municipal wastes were planned to treat by neutralization at first. Clarification was designed by followed neutralization employed on ash and local limestones in thickeners. The local pools could be beneficial later for humate anaerobic digestion on the quality of humate mud yield. However, all these process unıts were needed to control metal contents in the water paths and muddy wastes.

4.1. Thickening neutralization by fly ash

Acid mine drainage brings out great environmental issue of acidic metallic sulfate rich wastewaters. Those type wastewaters are also the by-products of a variety of industrial operations such as galvanizing, leaching, and the scrubbing of flue gases at thermal power plants (**Figure 8**). The major producer of such wastewaters is under the agricultural and environmental concern occurring in the coal and copper mining. In particular, abandoned mines seepage waters and copper mines wastewaters are often acidic. Those wastewaters cause an extra risk to the environment. They often contain high concentrations of metals (iron, lead, cadmium,

http://dx.doi.org/10.5772/intechopen.69927

Figure 8. The satellite image of MD clarified water pool in the Sırnak Coal Mine Site, fishery pool remediation.

zinc, aluminium, manganese, and rarely other heavy metals) and hazardous arsenic. AMD (**Figure 9**) waters damage as mine effluents. Although the true scale of the environmental pollution is caused by mine wastewaters, fly ash use on common discharges is difficult to assess accurately.

Acidic metal-rich waters may also form in spoil heaps and mineral tailings, essentially by the same biologically driven reactions as in mine shafts and adits. Due to the more disaggregated (and more concentrated, in the case of tailings) nature of the acid-generating minerals in these

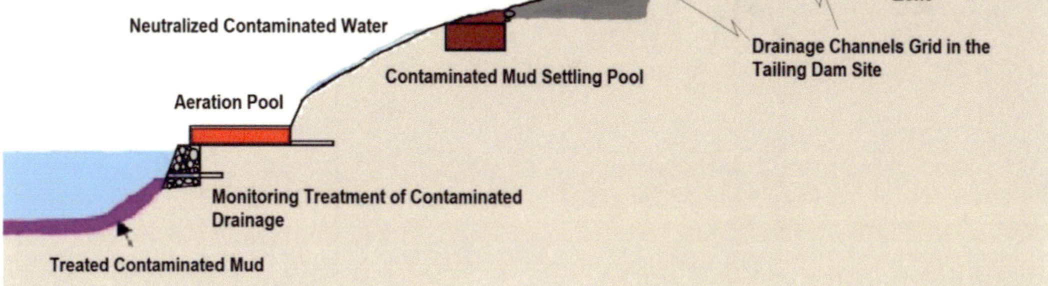

Figure 9. Acid mine drainage, contaminated water treatment by fly ash and neutralization by fly ash and mid-bottom ash of thermal power plants.

waste materials, the scanning electron microscopy pictures showed the Ca ferrites and active lime powders, which were effective in neutralizations.

AMD flowed from seepage channels was more aggressive than the discharges from the concentrator and ore stock site. Another critical concern was the potential of controlling the long-term contamination of agricultural irrigation water and soil contamination. The treatment of AMD waters in cycling manner between the slime ponds would continue for many years after the asphaltite extraction in mines stopped and tailing dams demolished. Although the neutralization pH of these waters was simple and above 6, particularly at the cycling depth (where dissolved oxygen concentrations was low), the metals such as the iron, zinc, and aluminium were dissolved and also the fly ash and bottom ash yielded local area as waste [9]. The ash reacted on neutralization caused reduced (Fe^{2+} and Mn^{2+}) ionic states in anoxic AMD waters, and these forms of the metals were much stable at higher pH than the fully oxidized ($Fe^{3+}$and Mn^{4+}) ions. The higher content of fly ash resulted in the AMD streams remained neutral-to-alkaline. Net acidity in AMD needs to be offset against any alkalinity of fly ash presented this was chiefly in the form of bicarbonate $(HCO_3)^-$ deriving from the dissolution of basic minerals (e.g., calcium carbonate), though, as noted below, biological processes may also generate alkalinity in AMD streams.

4.2. Thickener sedimentation

Fly ash slurries were collected in water pools and pumped to clarifiers for settlement and separation from alkali solutions as paste fine materials sedimentation and the settling of solid matter occur as below (**Figure 10**):

$$v_t = g\left(\frac{\rho_s - 1}{18\mu}\right)d_p^{\,2} \;\; Re \leq 2 \tag{14}$$

$$v_t = 0{,}13\left(g\left(\rho_s - 1\right)\right)^{0{,}72} d_p^{\,1{,}18} \mu^{0{,}45} \;\; 2 < Re < 500 \tag{15}$$

$$v_t = 1{,}74\left(g\left(\rho_s - 1\right)d_p\right)^{0{,}5} \;\; Re \geq 500 \tag{16}$$

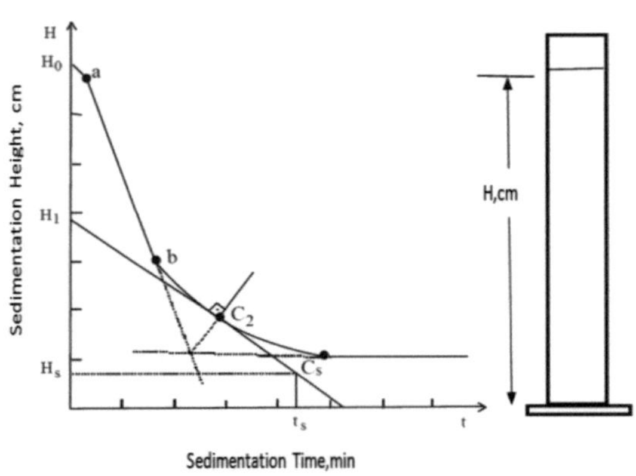

Figure 10. The settlement rate in thickner, sedimentation column used in settlement test.

http://dx.doi.org/10.5772/intechopen.69927

$$\mathfrak{R} = \frac{v_t d_p}{\mu} \quad (17)$$

Where v_t is the terminal settling velocity of the particle in thickener, g is the gravity acceleration, ρ_s is the specific gravity of settling particle in thickener, d_p is the diameter of settling particle in thickener, μ is the kinematic viscosity of fluid as water, $\mathfrak{R}$ is the Reynolds parameter of fluid flow as given in Eq. 12, and the sedimentation in laminal thickener occurs as given below as illustrated in **Figure 11**.

$$\mathrm{H} = \frac{Q}{A} * t_s \left(\frac{1}{\rho_l} + \frac{x}{\rho_s} \right) \quad (18)$$

H is the height of sedimentation of particles in thickener, t_s is the sedimentation time of settling particle in thickener, Q is the flow rate of solid matter in thickener, ρ_s is the specific gravity of settling particle in thickener, ρ_s is the specific gravity of liquid as water, x is the solid/liquid ratio in thickener feed.

Acid mine/rock drainage is a rampant problem in the world today. The screens illustrated in **Figure 10** and hydrolic clarifiers were tested (**Figure 11**) for the sediments in acid mine drainage treatment at an abandoned mine in the Şırnak. In this project, simple hydraulic clarifier may also provide neutralization reaction. It was very difficult for the government to enforce AMD treatment for the acid rock drainage at the Şırnak Coal Mine site. The conventional clarifier (**Figure 12**) and paste thickener (**Figure 13**) were also tested for fly ash neutralization.

Acid mine/rock drainage waters could be neutralized by fly ash and limestone waste fine in paste thickeners at longer times than conventional thickeners. The produced paste tailings occur reaction products and as mud as part of the chemical weathering process, but it is significantly used by mining, as filler material **Figure 14**.

AMD waters in Şırnak Coal Mine's lake created a sulfuric acid effluent. The sulfuric acid dissolved and leached out the metals that are in the site rocks and coal pyrite metals, creating

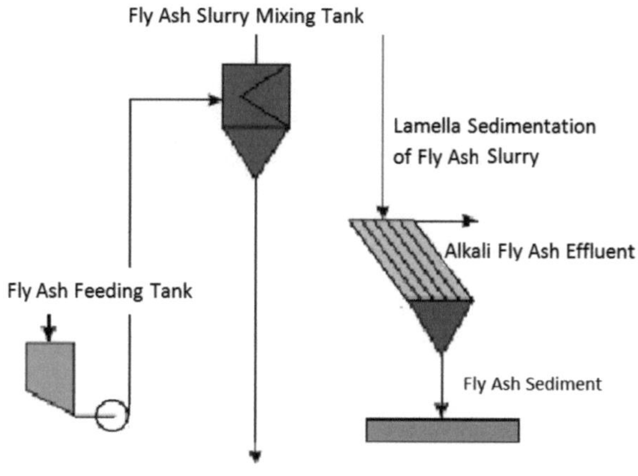

Figure 11. The lamella clarification of fly ah slurries in the Şırnak Coal mining AMD.

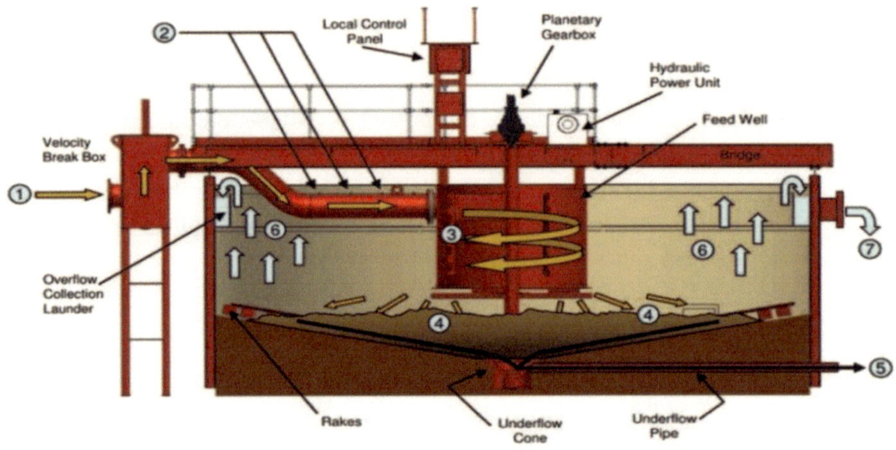

Figure 12. The WesTech thickener for fly ash and mid ash for clarification remediation in the Sırnak Coal Mine Site, fishery pool remediation.

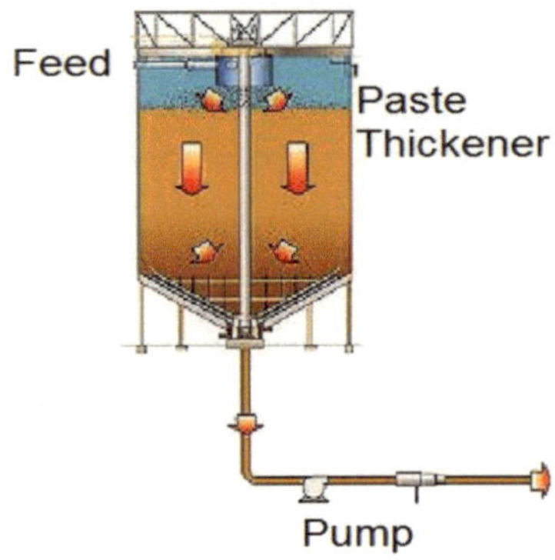

Figure 13. The Eimco paste thickener for fly ash and mid ash for clarification remediation.

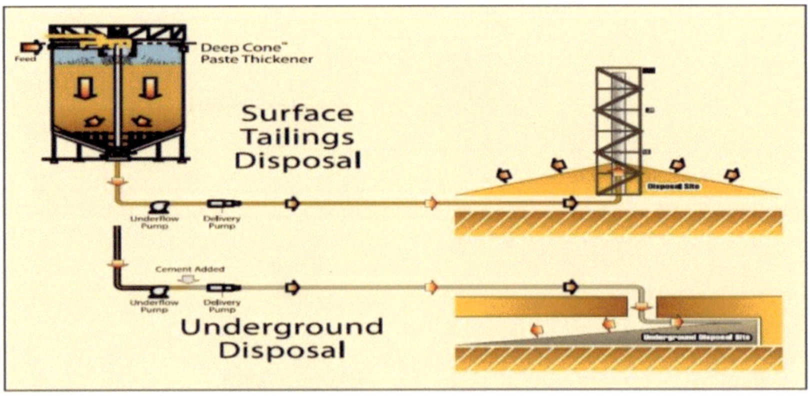

Figure 14. Paste thickening of treated neutralized effluent in the Coal mining AMD.

http://dx.doi.org/10.5772/intechopen.69927

high concentrations of dissolved metals such as iron, zinc, copper, arsenic, and cadmium. As the water is exposed to aeration and fly ash neutralization as changing amount of 10–100 kg/ton of effluent was effective as illustrated in **Figures 15** and **16**.

Acidic mine drainage waters were treated by fly ash and limestone in paste thickeners, and Şırnak fly ash and mid ash of thermal power station were used at equal weight rates between 10 and 100 kg/ton of AMD water. Some part of the rock weathering process was reacted very slow in the season of winter but in summer term, the reaction kinetics were higher, but the treatment neutralization occurred higher than in winter term due to temperature of AMD waters. Significantly, as water comes in contact with the fly ash and mid ash, it neutralized to pH 6. The neutralized waters contained dissolved metal contents such as Zn and Fe and effluents containing alkali K, and phosphate metals that are beneficial for organic soils. The unsettling muddy rust color of the water also served to make organic matter for agricultural

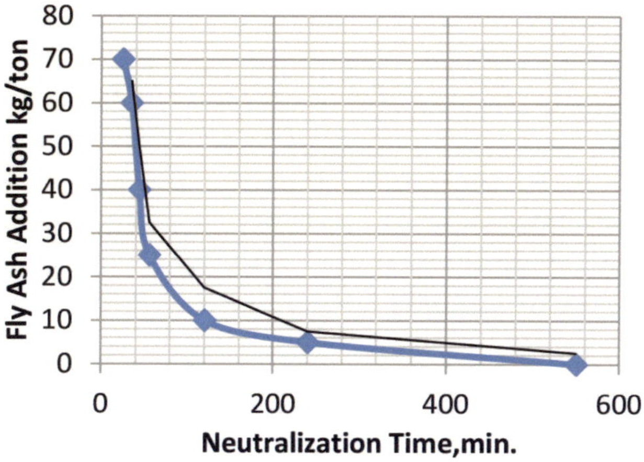

Figure 15. Effect of Fly Ash and Lime Use in Neutralization of AMD in the Şırnak Coal Mine Site.

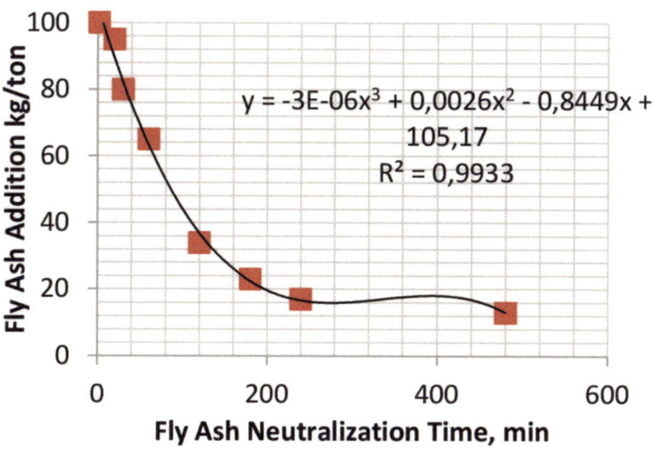

Figure 16. Effect of Fly Ash Use in Neutralization of AMD in the Şırnak Coal Mine Site.

irrigation waters in Şırnak Coal Mine site area. The settlement of muddy rust colored treated water occurred as illustrated in **Figure 16**. In brief, the major high-rate oxidation caused was the highly oxidation of iron pyrite (FeS_2) and other sulphidic minerals. As a consequence of the mining and processing of copper ores and coals, many metals (e.g., iron, zinc, and copper) were associated with sphalerite and pyrite in the Şırnak Coal Mining and Siirt Copper Mine. Additionally, coal deposits contain variable (generally, 1–20%) amounts of pyritic-sulfur (a generic term that includes other iron sulfide minerals such as marcasite) as well as organic sulfur. As seen in **Figure 17**, common neutralization treatment is carried out by limestone, lime, and caustic lime, but local wastes such as fly ash or mid-bottom ash of thermal power plants using desulfurization.

When the pyrite oxidized in coal tailings slimes and slurry in the dumping area and slime pond in Şırnak, lack of oxygen occurred in the pond, making the water contaminated for farming and aquatic life. Moreover, the rust and heavy metals in the water make it impossible life limiting fishing [51–64].

The regeneration of ferric iron was the key reaction in promoting acidic media contamination during oxidation of the pyrite mineral in asphaltite slime pond at pH values above 4. This commenced mediated chemically or biologically, as below pH 5, abiotic iron oxidation is negligible [1–5, 65–75], and the activities of moderately and extremely acidophilic iron-oxidizing bacteria had a main role with a lack of oxygen in the slime pond. The contaminated water cycled by the neutralization unit to pond for treatment of acid mine drainage seepage. The sand separation followed neutralization was applied by Derrick classifiers and aeration of

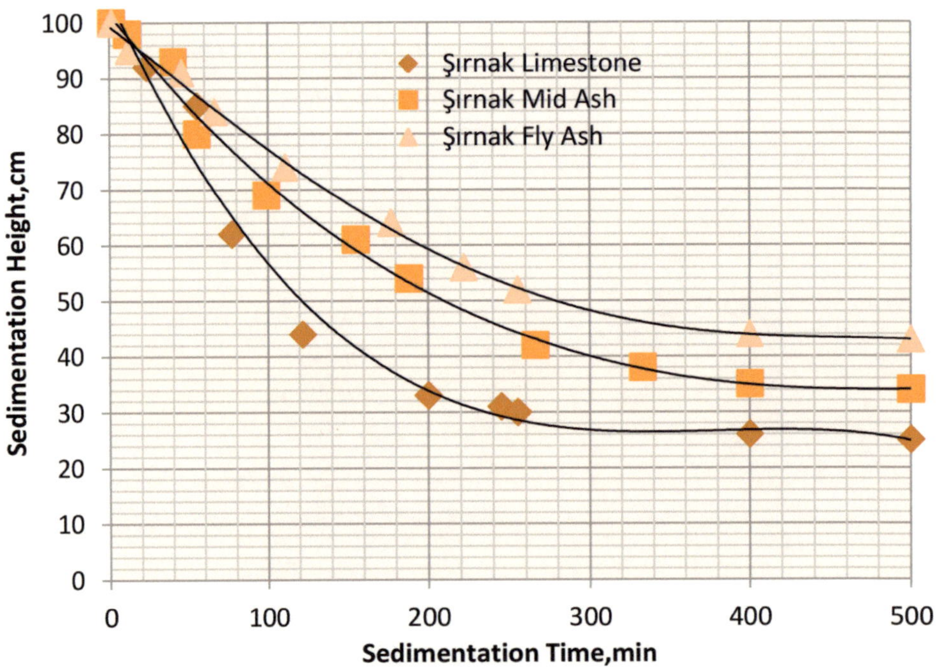

Figure 17. Effect of Fly Ash Use in Sedimentation of AMD in the Şırnak Coal Mine Site.

http://dx.doi.org/10.5772/intechopen.69927

wastewaters carried out in the Eriez floaters and thickeners as illustrated in **Figures 18** and **19** [61, 62].

Acid mine drainage (AMD) neutralized by flay ash and alkali matters followed the formation settled mud in the classifiers in cycling slime pond in coal mine. Although this is generally of humate mud yield was also circulated to collect the mud in the slime pond by asphaltite slime had an active on fertilizer production by carbon and water tables are kept artificially low by pumping. This initial drainage water tends to be more potentially polluting (in terms of acidity and metal content) than AMD that is discharged subsequently [22, 76–77].

In this treatment, study of coal mine AMD treatment by use of fly ash and mid ash managed the neutralization by sediments such as jarosites and gypsum depending on oxygen content. The precipitation of jarosite and göthite clusters observed as given in the Eqs. 19 and 20 showed the reactions occurred between pH 3 and 6. As illustrated in **Figures 20** and **21**, the alkali reactions followed the paths over the arrows down till the lack of oxygen side at the bottom of wastewaters level. The alkali pH created much precipitation and jelatizing of sediments and deteriorating solid/liquid neutralization reactions in the thickeners.

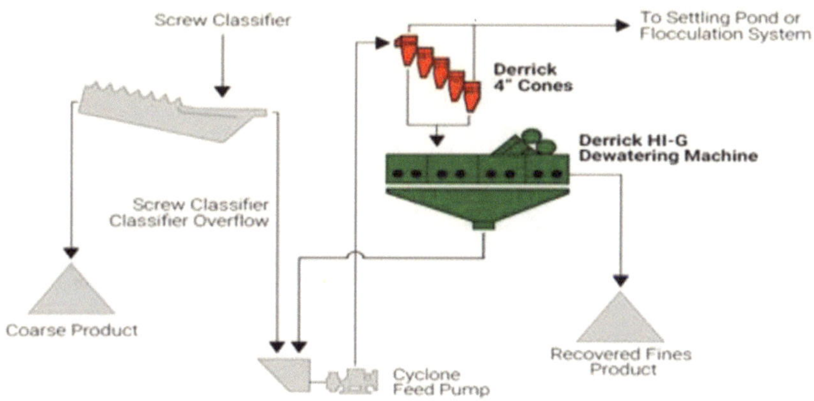

Figure 18. The Derrick classifier for mud humate yield for clarification remediation [61].

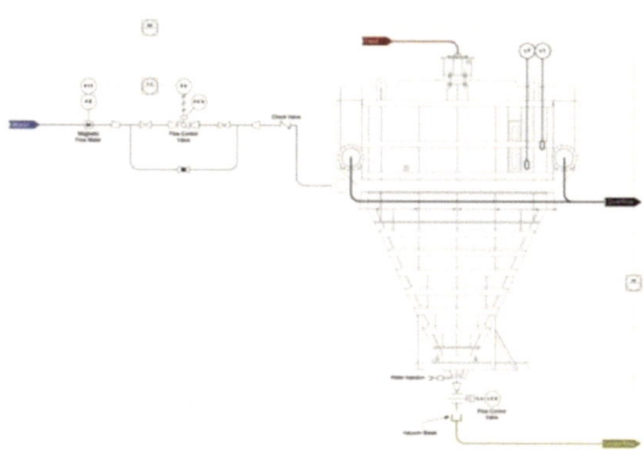

Figure 19. The classifier for mud digestion for AMD treatment in the Şırnak [62].

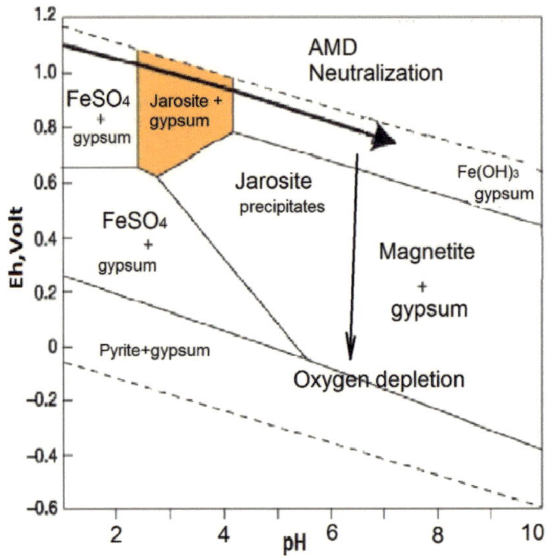

Figure 20. The Eh-pH phase diagrams in the Fe-S-O-H and Cu-S-O-H system.

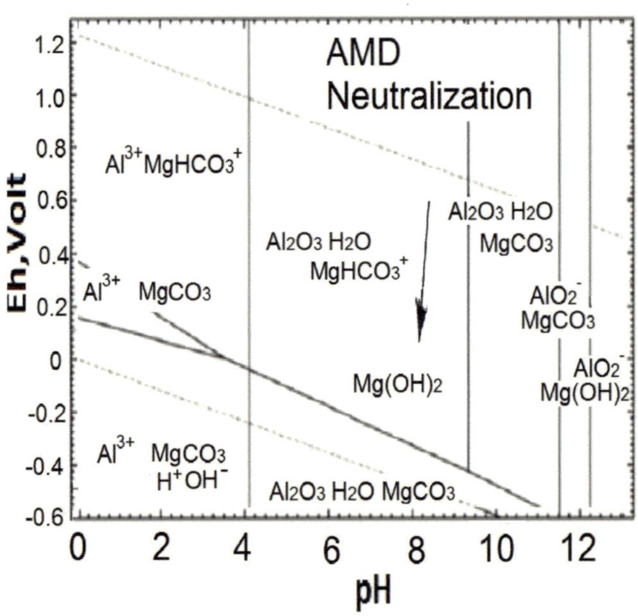

Figure 21. The Eh-pH phase diagrams with the fly ash clay and Al, Mg-OH, and CO_2-OH system.

4.3. Humate mud yield and soil remediation

In this treatment, study of coal mine AMD treatment by use of fly ash and mid-bottom ash managed the neutralization by sediments and solids were analyzed by ASTM Soil Test standards as given in **Table 2**. The results of treated wastewater muds are given in **Table 3**.

Analysis	Method
pH	Water
Lime requirement	Mehlich buffer
Extractable P, K, Ca, and Mg	Mehlich 3 (ICP)
Cation Exchange Capacity (CEC)	Summation
Organic matter	Loss on Ignition
Total C	Combustion
Nitrate N	Specific Ion Electrode
Ammonium N	Specific Ion Electrode
Total N	Combustion
Soluble salts	Electrical Conductivity (1:2)
Total sorbed Cu, Zn, Pb, Ni, Cd, Cr, Mo	EPA Method 3050B/3051 + 6010
Particle size analysis	Hydrometer method
As	EPA Method 3050B/3051 + 6010
Se	EPA Method 3050B/3051 + 6010
Hg	EPA Method 7473
Calcium carbonate equivalency (CCE)	ASTM Method C 25

Table 2. List of the standard soil-testing methods used soil remediations.

Humate waste sludge was treated by densification and CO_2 gas or biogas digestion in the concrete column pools sequentially with the flow sheet as given in **Figure 22**.

The plots treated with poultry manure (N-based) and TSP, without digestion amendment, had DPS values of 0.84 and 0.93, respectively, as given in **Table 3** in which these treatments could contribute humate content with significant amounts of P in runoff by anaerobic digestion by own biogas. The kinetic rate of digestion was greatly depended on active sludge carbon content and temperature. During digestion of a month, period time might not sufficient for fertile quality of mud, and organic manure carbon matter in 0.1% alkaline. Amendment with digestion and manure decreased the N and the TSP values by 86% (N-based) and 82% (P-based) relative to the same treatments without digestion amendment, respectively. Similarly, N and DPS values of the manure-treated mud decreased by 83% (N-based) and 82% (P-based), respectively.

$$4Fe^{3+},\ 2Zn^{2+},\ 2Cu^{2+} + 14\ O_2 + 7\ H_2O \rightarrow 4Fe(OH)_3 + 2Cu(OH)_2 4Zn(OH)_2 + 8SO_4{}^{2-} + 6H^+ \tag{19}$$

$$4Na^+, 4K^+, 2Ca^{2+} 14O_2 + 7H_2O + 8SO_4{}^{2-} \rightarrow 2NaOH + 2KOH + 2CaSO_4 + Na_2/K_2(SO_4)(OH)_6 Fe(OH)_3 \tag{20}$$

The metal contents were higher with treatment of fly ash and bottom ash use in neutralization muds. The low power requirements of aeration limestone channel treatment made much

	Bottom ash + 50% soil	Manure	Humate	Fly ash + 50% Soil	Avgamasya soil	AMD treated limestone	AMD bottom ash
pH	6.8	8.2	7.9	5.6	6.8	7.2	7.9
Total C	32.0	54.7	56.6	13.9	34.7	16.6	13.9
Total N	0.4	7.0	9.03	0.7	8.0	1.03	0.7
% Solids	27.0 + 4.2	13.4 + 1.6	15.4 + 3.4	40.6 + 6.2	15.4 + 3.4	40.6 + 6.2	40.6 + 6.2
Total P	18.9 + 3.8	38.7 + 2.3	24.1 + 4.2	4.69 + 0.7	24.1 + 4.2	4.69 + 0.7	4.69 + 0.7
Total Al	0.94 + 0.1	9.37 + 0.4	9.26 + 1.4	107 + 8.3	9.26 + 1.4	107 + 8.3	107 + 8.3
Total Fe	1.53 + 0.3	24.3 + 3.1	32.8 + 6.4	6.08 + 0.4	32.8 + 6.4	6.08 + 0.4	6.08 + 0.4
Oxalate P	10.4 + 1.2	26.4 + 3.9	28.5 + 3.2	4.33 + 0.8	28.5 + 3.2	4.33 + 0.8	4.33 + 0.8
Oxalate Al	0.79 + 0.1	6.50 + 0.9	7.41 + 0.6	84.3 + 6.2	7.41 + 0.6	84.3 + 6.2	84.3 + 6.2
Oxalate Fe	10.82 + 3.3	19.4 + 2.3	24.7 + 4.2	5.16 + 1.0	3.82 + 1.3	4.22 + 1.3	5.16 + 1.0
Fe strip P	1.34 + 0.3	6.42 + 1.2	1.05 + 0.2	nd[d]	1.05 + 0.2	nd[d]	nd[d]
WEP[a]	0.85 + 0.1	2.59 + 0.6	0.34 + 0.1	nd[d]	0.85 + 0.1	nd[d]	nd[d]
PSI[b]	22	1.44	0.7	33	11	6	43

[a] Water extractable P.
[b] Phosphorus saturation index.
[c] Not applicable.
[d] Not determined.

Table 3. Selected chemical properties of remediation product sediment muds and soils in the Sırnak Coal Mine Site.

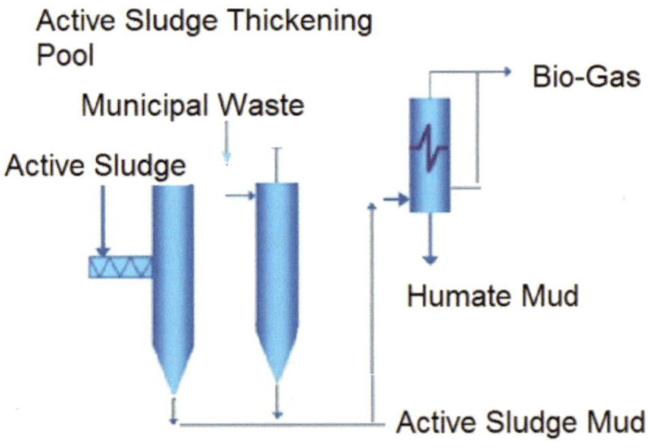

Figure 22. Selected chemical digestion of mud for remediation product sediment muds and soils in the Şırnak Coal Mine Site.

feasible neutralization for disseminated asphaltite extraction and slimes in many locations of Şırnak. The distribution of asphaltite slime ponds allowed to cycling at low flow rates in neutralization and required less cost. However, the pool settlers had a lower capacity and could not remove any of the heavy metals in the cycling itself. For concentrator tailing ponds, the larger flows in cycling manner removed the heavy metals by filtration treatment system. It was necessary to employ a more active and filtration application for heavy metal contents followed thickening treatment system so the humate mud contained less dissolved metal contamination.

5. Projected flow sheet

Şırnak Coal Mine drainage water was not always strictly acidic. The projected neutralization plant to develop treatment for neutral mine drainage water that contained a fairly normal pH around 6 but contained high amount of heavy metals at the concentrations in the cycling slime waters over 200 mg/lt Fe and over 45 mg/lt Zn higher Pb and Cd and other contaminants that needed to be removed in order to farming requirements in this project. In the asphaltite slime water cycling with less acidic, the use of lime was not required to settle out the iron. In heavy dense settling and neutralization process, hydrogen peroxide or caustic lime could be concerned potentially use, but locally cheap alkali was concerned. However, without the lime and fly ash, the thickener bed was bound higher, and settling rate was decreased relatively as low as 5% solids. The proposed design flow sheet is illustrated in **Figure 20** on the fly ash content paste thickening proposed on the solid content in wastewater flow. In the copper mine site of Siirt, there was necessary to use collector pools and paste thickeners due to high solid content over 13%. In fact, in many cases, the thickener itself might not be necessary. The projected design considered the use of a recirculating-type clarifier or a solid contact clarifier.

For large acid mine flows, the projected fly ash treatment has developed as the following flow sheet proposed by this study shown in **Figure 23**. As an equipment manufacturer specializing in liquid/solids separation, WesTech has designed and installed many acid mine drainage treatment systems. Lack of digestion at a month caused wet acidic mud, and organic fumic matter was lower than 20% in 1% K-alkali test. Amendment with digestion and manure addition at 2 months' period increased the N and TSP by 86% (N-based) and 82% (P-based) relative to the same treatments at 1 month digestion amendment, respectively. The metal contents of humate mud were critical for organic farming and agricultural irrigation practice in the region.

37 mg/lt Cu, 52 mg/lt Zn, and about 141 mg/lt Fe concentration as wastewater contaminants caused by acidic mine drainage waters of coal mine extraction site due to wet soils and environment and stream water channels determined as concentrated and polluted load. The pollutant load and pollutant concentrations changed widely and treated in the channel streams by fly ash and limestones used for various purposes. The thickening capacity was designed by flow of 2–3 lt/s wastewater stream at 4–7% solid contents. The design capacity of the wastewater treatment plant resulted the lower metal contents in water analysis as clarified waters, but lead and zinc levels were still higher for fishing. The agricultural irrigation was managed by this water. The thickening output water oxygen demands were net below 10 and 10 COD and BOD levels (BOD, kg/day) and at the digestion, carbon dioxide and biogas entrance to the digestion treatment plant were expressed by the mud pollution load to be found as the product of the multiplication. The design capacity depends on the number of sequential kinetic rates of the mud content and temperature at the digestion pool depending on the pool conditions. The pollution loads of return currents were greatly affected much. Mud

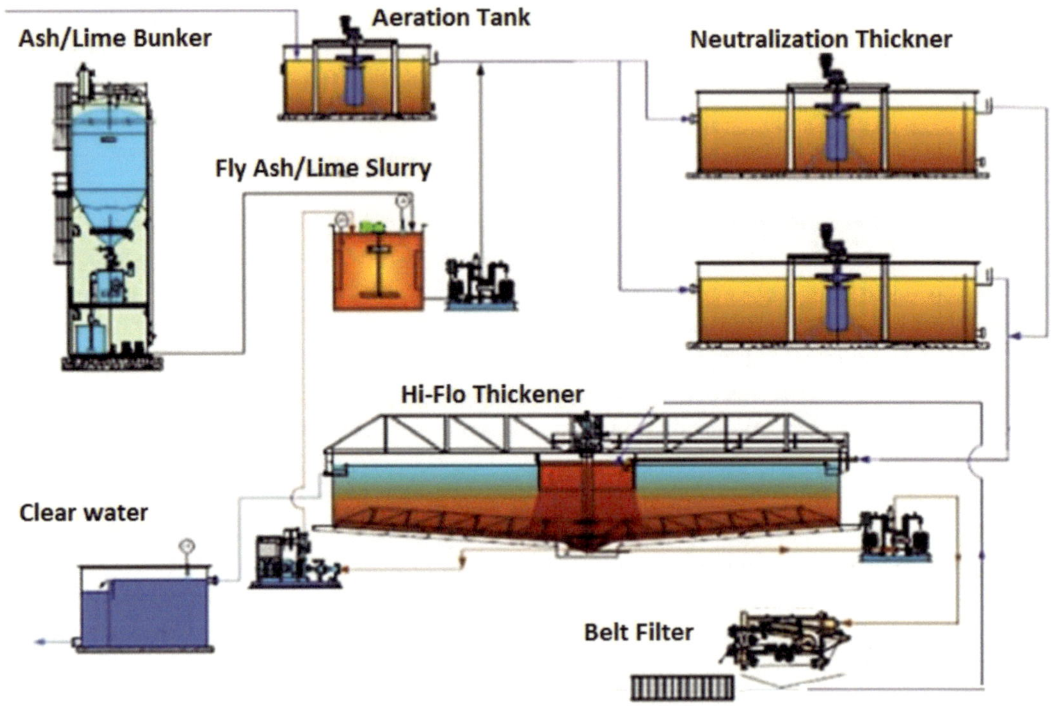

Figure 23. Projected Neutralization and Treatment Plant for AMD waters in Şırnak Coal Mine Site.

quality increased at humate value of organic dissolution, and phosphor level was measured at dry seasons with 30 BOD level on the summer times at temperatures over 27°C.

Neutralization and settling step in the treatment process managed the water clarification and removing the nonsoluble iron from the liquid. Heavy metal contents in the clarified stream decreased by ideal neutralization, allowing fly ash use to slow the settling rates down as necessary neutralization time. It is obviously critical to add neutralization ultrafine waste ash or fly ash into the HDS clarifier.

6. Conclusions

In this area, the hydrological characteristics of the Şırnak city were critical for the effect of waste characteristics on the pictured coal site and city. The hydrological features of the urbanization and countryside for agricultural irrigation with new settlements need modeling regarding available water source. The hydrological property of settlement areas with dense populated areas should be determined by geographic information systems (GIS) techniques. The main purpose of this municipal wastewaters and acidic waters treatment was to investigate the effect on settlement on the basic hydrological structure by studying the characteristics of the ground topography, ground water elevation, slope, and viewing. GIS techniques were used in the creation of the thematic maps and in the analysis of the parameters. The GIS study models created, the available water source changed, and a stream network model was provided sufficient source control at the Şırnak province. GIS data maps followed planning of wastewater treatment plant made it possible to obtain more qualified data by enabling the use of humate mud production regarding metal contents in this research.

The AMD waters during the coal mining in the open pits and underground in Şırnak and copper AMD waters in Siirt copper mining and concentrator are polluting nature. This type of wastewater is acidic and contains heavy metals, which cause environmental pollution such as ferric iron and zinc and copper metals at 230, 122, and 57 mg/lt, respectively. For this reason, the release of acidic mine drainage (AMD) waters into the irrigation waters of natural rivers and agricultural areas can cause great threat to the habitats and fish in the local streams and in the local lakes. AMD waters must be collected in the impermeable projected ponds and require neutral characterization and ventilation by acidic neutralization.

A more useful treatment method projected using local limestones, and fly and mid-bottom ash of thermal power stations used for desulfurization was so effective among the known remediation technologies that rely on biological activities and produced beneficial potassium and phosphate fertilizing mud.

Within these major treatments, there are processes that may be described as either active or passive neutralization by fly ash and mid-bottom ash with limestones of Şırnak at equal rates of 60 kg/lt AMD water. This choice projected in Şırnak coal mining and Siirt copper mining site could remediate AMD waters following neutralization at economical and environmental stages.

The true environmental cost of this remediation system was low at apparent study with investment cost of about $2 million. One of limiting operational costs was the amount of the transport of liming materials, often low distances from source to mine sites with a $21 per ton alkali. In the Siirt copper open pit mining site in Sirvan, large acidic seepage volumes as tailing wastewaters could be treated by active chemical reactions of neutralization, made low acidic or neutral waters. The land surface area and topography brought out passive remediation systems in Siirt copper mine. However, similar treatment in the coal mining in Şırnak became significant at high costs of lime addition and treatment of sludge disposal. The higher land areas required for passive treatments made dramatically at smaller area by densing on small collector ponds optimizing biological processes, for example, packed bed bioreactors for removing iron from acidic mine waters following discharge, which were effective than aerobic wetlands [3, 9].

The projected passive pool systems also required a certain amount of management and would eventually fill with accumulated ferric mud (aerobic wetlands) and fertile muds. The long-term stabilities of these materials were certain, but since as the composted effluents in some pools might contain toxic elements (arsenic and cadmium such as 10 mg/lt and 56 mg/lt, respectively), their storage or disposal required careful consideration.

The sustainability of the projected remediation system was a great fact that was becoming increasingly critical in decision making. One of the problems here was that the products of AMD remediation have not been perceived as a resource. Recently, an iron oxide sludge recovered from a drainage channel at an abandoned coal mine in Şırnak coal site could be used to manufacture fertilize mud in a commercially packed bag.

The Cu and Zn metals could also be recovered by electrolysis as by products from AMD waters [9]. The treatment of AMD from metal mines in Siirt copper concentrator could provide some financial return on the investment and running costs of remediation. Ultimately, legislation was a prominent fact to become the dominant factor in determining which kind of remediation method could be used in local needs. The projected thickener neutralization of pond effluents might become increasingly beneficial to dispose of the base metals in sludges and sediments avoiding storage problems. There were high possibilities for their recovery and recycling.

The ecological limits for agricultural irrigation waters of wheat in Şırnak at the 50 mg/lt concentrations of metal sulfate that could be discharged from treatment plants needed the choice of two-stage systems as one that effectively removes sulfate as well as metals and acidity from mine waters, and the other stage may decrease the metal concentrations by water and effluent mixing. The problem of what to do about the pollution threat posed by AMD certainly brings to a certain treatment using fly ash and waste limestone forever in coal and copper mining activities at environmental concern.

Author details

Yıldırım İsmail Tosun

Address all correspondence to: yildirimismailtosun@gmail.com

Mining Engineering Department, Şırnak University, Şırnak, Turkey

References

[1] Johnson DB, Hallberg KB. Pitfalls of passive mine water treatment. Reviews in Environmental Science and Bio/Technology. 2002;**1**:335-343

[2] Johnson DB, Hallberg KB. The microbiology of acidic mine waters. Research in Microbiology. 2003;**154**:466-473

[3] Johnson DB. Biological removal of sulfurous compounds from inorganic wastewaters. In: Lens P, Hulshoff Pol L, editors. Environmental Technologies to Treat Sulfur Pollution: Principles and Engineering. London: International Association on Water Quality; 2000. pp. 175-206

[4] Johnson DB. Chemical and microbiological characteristics of mineral spoils and drainage waters at abandoned coal and metal mines. Water, Air, & Soil Pollution: Focus. 2003;**3**:47-66

[5] Johnson DB, Hallberg KB. Acid mine drainage remediation options: A review. Science of the Total Environment. 2005;**338**(1–2):3-14

[6] Kalin M, Cairns J, McCready R. Ecological engineering methods for acid-mine drainage treatment of coal wastes. Resources, Conservation and Recycling. 1991;**5**:265-275

[7] Kalin M, Chaves WLC. Acid reduction using microbiology (ARUM) treating AMD effluent emerging from an abandoned mine portal. In: Ciminelli VST, Garcia Jr O, editors. Biohydrometallurgy: Fundamentals, Technology and Sustainable Development. Vol. 11B. Amsterdam: Elsevier; 2001. pp. 289-296

[8] Kalin M, Fyson A, Wheeler WN. The chemistry of conventional and alternative systems for the neutralization of acid mine drainage. The Science of the Total Environment. 2006;**366**(2–3):395-408

[9] Tosun YI. 2017AMD Neutralization by Bottom Fly Ash - Remediation of Agricultural Fields by Organic Humate Mud of AMD Treatment, New Trends in Mining, Proceedings of IMCET 2017, 25thTurkish Mining Congress, TMMOB, Antalya, Turkish, pp. 804-812

[10] Li MG, Aube BC, St-Arnaud LC. Considerations in the use of shallow water covers for decommissioning reactive tailings. In: Proceedings of the Fourth International Conference on Acid Rock Drainage; May 30–June 6, 1997; Vancouver, BC. Vol. I; 1997. pp. 115-30

[11] Landa ER. Microbial biogeochemistry of uranium mill tailings. Advances in Applied Microbiology. 2005;**57**:113-130

[12] Ledin M, Pedersen K. The environmental impact of mine wastes - Roles of microorganisms and their significance in treatment of mine wastes. Earth-Science Reviews. 1996;**41**: 67-108

[13] Long ZE, Huang YH, Cai ZL, Cong W, Fan OY. Biooxidation of ferrous iron by immobilized Acidithiobacillus ferrooxidans in poly(vinyl alcohol) cryogel carriers. Biotechnology Letters. 2003;**25**:245-249

[14] Silva LFO, Fdez-Ortiz de Vallejuelo S, Martínez-Arkarazo I, Castro K, Madariaga JM. Study of environmental pollution and mineralogical characterization of sediment rivers from Brazilian coal mining acid drainage. Science of the Total Environment. 2013;**447**:169-178

[15] Stumm W, Morgan JJ. Aquatic Chemistry: An Introduction Emphasizing Chemical Equilibria in Natural Waters. NY: Wiley-Interscience; 1981

[16] Mehling PE, Day SJ, Sexsmith KS. Blending and layering wasterock to delay, mitigate or prevent acid generation: A case review study. In: Proceedings of the Fourth International Conference on Acid Rock Drainage; May 30–June 6, 1997; Vancouver, BC. Vol. II; 1997. pp. 953-970

[17] Millero FJ, Pierrot D. A chemical equilibrium model for natural waters. Aquatic Geochemistry. 1998;**4**(1):153-199

[18] Evangelou VP. Pyrite chemistry: The key for abatement of acid mine drainage. In: Geller A, Klapper H, Salomons W, editors. Acidic Mining Lakes: Acid Mine Drainage, Limnology and Reclamation. Berlin: Springer; 1998. pp. 197-222

[19] Loos MA, Bosch C, Mare' J, Immelman E, Sanderson RD. Evaluation of sodium lauryl sulfate, sodium benzoate and sorbic acid as inhibitors of acidification of South African coal waste. In: Groundwater and Mining: Proceedings of the 5th Biennial Symposium of

the Groundwater Division of the Geological Survey of South Africa Randberg, Transvaal. Pretoria: Geological Society of South Africa; 1989. pp. 193-200

[20] Brown M, Barley B, Wood H. Minewater Treatment-Technology, Application and Policy. London, UK: IWA Publishing; 2002

[21] Christov C, Moller N. A chemical equilibrium model of solution behavior and solubility in the H-Na-K-Ca-OH-Cl-HSO4-SO4-H2O system to high concentration and temperature. Geochimica et Cosmochimica Acta. 2004;**68**(18):3717-3739

[22] Clarke LB. Coal Mining and Water Quality. London: IEA Coal Research; 1995. pp. 99

[23] Coulton R, Bullen C, Dolan J, Hallet C, Wright J, Marsden C. Wheal Jane mine water active treatment plant-design, construction and operation. Land Contamination and Reclamation. 2003;**11**:245-252

[24] Coulton R, Bullen C, Hallet C. The design and optimization of active mine water treatment plants. Land Contamination and Reclamation. 2003;**11**:273-280

[25] Coupland K, Battaglia-Brunet F, Hallberg KB, Dictor MC, Garrido F, Johnson DB. Oxidation of iron, sulfur and arsenic in mine waters and mine wastes: An important role for novel Thiomonas spp. In: Biohydrometallurgy: A Sustainable Technology in Evolution: Proceedings on the 15th International Biohydrometallurgy Symposium; 2003

[26] Cruz R, Monroy M. Evaluation of the reactivity of iron sulfides and mining wastes. A methodology based on the application of cyclic voltammetry. Química Nova. 2006;**29**(3): 510-519

[27] Dayton EA, Basta NT, Jakober CA, Hattey JA. Using treatment residuals to reduce phosphorus in agricultural runoff. Journal of the American Water Works Association. 2003;**95**:151-158

[28] Klapper H, Geller W. Water quality management of mining lakes- a new field of applied hydrobiology. Acta Hydrochimica et Hydrobiologica. 2002;**29**:363-374

[29] Kleinmann RLP, Hedin RS, Nairn RW. Treatment of mine drainage by anoxic limestone drains and constructed wetlands. In: Geller A, Klapper H, Salomons W, editors. Acidic Mining Lakes: Acid Mine Drainage, Limnology and Reclamation. Berlin: Springer; 1998. pp. 303-319

[30] Kleinmann R. At-source control of acid mine drainage. International Journal of Mine Water. 1990;**9**(1–4):85-96

[31] Kontopoulos A. Acid mine drainage control. In: Castro SH, Vergara F, Sánchez MA, editors. Effluent Treatment in the Mining Industry. University of Concepción. 1998. pp. 57-118

[32] Kuyucak N. Effective passive water treatment systems in extremely cold Canadian climatic conditions. In: SME Annual Meeting and Exhibit; 2010. pp. 50-54

[33] Benner SG, Blowes DW, Ptacek CJ. A full-scale porous reactive wall for prevention of acid mine drainage. Ground Water Monitoring and Remediation. 1997;**17**:99-107

[34] Blodau C, Peiffer S. Thermodynamics and organic matter: Constraints on neutralization processes in sediments of highly acidic waters. Applied Geochemistry. 2003;**18**:25-36. http://dx.doi.org/10.1016/S0883-2927(02)00052-56

[35] Blowes DW, Ptacek CJ, Benner SG, McRae CWT, Bennett TA, Puls RW. Treatment of inorganic contaminants using permeable reactive barriers. Journal of Contaminant Hydrology. 2000;**45**:121-137. http://dx.doi.org/10.1016/S0169-7722(00)00122-4

[36] Boonstra J, van Lier R, Janssen G, Dijkman H, Buisman CJN. Biological treatment of acid mine drainage. In: Amils R, Ballester A, editors. Biohydrometallurgy and the Environment Toward the Mining of the 21st Century. Vol. 9B. Amsterdam: Elsevier; 1999. pp. 559-567

[37] Bornstein J, Hedstrom WE, Scott FR. Oxygen Diffusion Rate Relationships under Three Soil Conditions. Life Sciences and Agriculture Experiment Station Technical Bulletin 98; 1980

[38] Elliott HA, O'Connor GA, Brinton S. Phosphorus leaching from biosolids amended sandy soils. Journal of Environmental Quality. 2002;**31**:1362-1369

[39] Elliott HA, O'Connor GA, Lu P, Brinton S. Influence of water treatment residuals on phosphorus solubility and leaching. Journal of Environmental Quality. 2002;**31**:681-689

[40] Hallberg KB, Johnson DB. Biodiversity of acidophilic microorganisms. Advances in Applied Microbiology. 2001;**49**:37-84

[41] Hedin RS. Recovery of marketable iron oxide from mine drainage in the USA. Land Contamination and Reclamation. 2003;**11**:93-97

[42] Himsley A, Bennett JA. New continuous packed-bed ion exchange system applied to treatment of mine water. Ion Exchange Technology. 1984;144-152

[43] Bott TL, Jackson JK, Mctammany ME, Newbold JD, Rier ST, Sweeney BW, Battle JM. Abandoned coal mine drainage and its remediation: Impacts on stream ecosystem structure and function. Ecological Applications. 2012;**22**(8):2144-2163

[44] Brodie GA. Staged, aerobic constructed wetlands to treat acid drainage: Case history of Fabius Impoundment 1 and overview of the Tennessee Valley Authority's program. In: Moshiri GA, editor. Constructed Wetlands for Water Quality Improvement, Boca Raton, FL: Lewis Publishers; 1993. pp. 157-165

[45] Broman PG, Haglund P, Mattson E. Use of sludge for sealing purposes in dry covers-development and field experiences. In: Proceeding 2nd International Conference Abatement of Acidic Drainage; 1991; Ottawa, Canada; 1991. pp. 515-527

[46] Agyin-Birikorang S, O'Connor G, Oladeji O, Obreza T, Capece J. Drinking-water treatment residual effects on the phosphorus status of field soils amended with biosolids, manure, and fertilizer. Communications in Soil Science and Plant Analysis, 2008;1700-1719. http://dx.doi.org/10.1080/00103620802073644

[47] Aube BC, Payant S. The Geco process: A new high density sludge treatment for acid mine drainage. In: Proceedings of the Fourth International Conference on Acid Rock Drainage; May 30–June 6, 1997; Vancouver, BC. Vol. I. 1997. pp. 165-180

[48] Doupé RG, Lymbery AJ. Environmental risks associated with beneficial end uses of mine lakes in southwestern Australia. Mine Water and the Environment. 2005;**24**(3):134-138

[49] Eghball B, Power JL. Composted and noncomposted manure application to conventional and no-tillage systems: Corn yield and nitrogen uptake. Agronomy Journal. 1999;**91**:819-825

[50] Egiebor NO, Oni B. Acid rock drainage formation and treatment: A review. Asia-Pacific Journal of Chemical Engineering. 2007;**2**(1):47-62

[51] Umrania VV. Role of acidothermophilic autrotrophs in bioleaching of mineral sulphides ores. Indian Journal of Biotechnology. 2003;**2**(3):451-464

[52] Valiente M, Diez S, Masana A, Frías C, Muhammed M. Separation of copper and zinc from waste acidic mine effluents of Río Tinto area. Mine Water and the Environment. 1991;**10**(1):17-27

[53] Wiegand U, Schreck P, Schreiter P, Lerche I, Glaesser W. Restoration of open pit lignite mining in the former GDR: Lessons to be learnt from Zwenkau. Energy and Environment. 2003;**14**(4):437-449

[54] Younger PL, Coulton RH, Froggatt EC. The contribution of science to risk-based decision-making: Lessons from the development of full-scale treatment measures for acid mine waters at Wheal Jane, UK. Science of the Total Environment. 2005;**338**:137-154

[55] Neal C, Whitehead PG, Jeffery H, Neal M. The water quality of the River Carnon, west Cornwall, November 1992 to March 1994: The impacts of Wheal Jane discharges. Science of the Total Environment

[56] Neculita CM, Zagury GJ, Bussière B. Passive treatment of acid mine drainage in bioreactors using sulfate-reducing bacteria: Critical review and research needs. Journal of Environmental Quality. 2007;**36**:1-16

[57] Nordstrom DK. Mine waters: Acidic to circumneutral. Elements. 2011;**7**(6):393-398

[58] Palmer C, Young PJ. Protecting water resources from the effects of landfill sites: Foxhall landfill site. Water and Environmental. 1991;**5**(6):682-696

[59] Ackman T, Erickson P. In-Line Aeration and Neutralization (ILS)-Summary of Eight Field Tests. In: AIME/SME/TMS 115th Annual Meeting. New Orleans, LA. 1986. p. 34

[60] Battaglia-Brunet F, Dictor MC, Garrido F, Crouzet C, Morin D, Dekeyser K, et al. An arsenic(III)-oxidizing bacterial population: Selection, characterization, and performance in reactors. Journal of Applied Microbiology. 2002;**93**:656-667

[61] Anonymous. (Eriez Co) [Internet]. 2016. Available from: https://www.eriezflotation.com/flotation/crossflow-classifier

[62] Anonymous. (Derrick Co) [Internet]. 2016. Available from: https://www.derrick.com/Hi-G/

[63] Beauchemin S, Fiset JF, Poirier G, Ablett JM. Arsenic in an alkaline AMD treatment sludge: Characterization and stability under prolonged anoxic conditions. Applied Geochemistry. 2012;**25**(10):1487-1499

[64] Evangelou VP. Pyrite Oxidation and its Control. New York: CRC Press; 1995. pp. 275

[65] Fernández-Rubio R, Lorca DF. Mine water drainage. Mine Water and the Environment. 1993;**12**:107-130

[66] Grande JA. Impact of AMD processes on the public water supply: Hydrochemical variations and application of a classification model to a river in the Iberian Pyritic Belt, SW Spain. Hydrology Research. 2011;**42**(6):472-478

[67] Kepler DA, McCleary EC. Successive alkalinity producing systems (SAPS) for the treatment of acidic mine drainage. In: Proceedings of the International Land Reclamation and Mine Drainage Conference and the 3rd International Conference on the Abatement of Acidic Drainage; April, 1994; Pittsburgh, PA. Vol. 1; 1994. pp. 195-204

[68] McCullough CD, Lund MA. Opportunities for sustainable mining pit lakes in Australia. Mine Water and the Environment. 2006;**25**:220-226

[69] McDonald DM, Webb JA, Taylor J. Chemical stability of acid rock drainage treatment sludge and implications for sludge management. Environmental Science & Technology. 2006;**40**(6):1984-1990

[70] Molson JW, Fala O, Aubertin M, Bussière B. Numerical simulations of pyrite oxidation and acid mine drainage in unsaturated waste rock piles. Journal of Contaminant Hydrology. 2005;**78**(4):343-371

[71] Monterroso C, Macías F. Drainage waters affected by pyrite oxidation in a coal mine in Galicia (NW Spain): Composition and mineral stability. Science of the Total Environment. 1998;**216**(1):121-132

[72] Nancucheo I, Johnson DB. Selective removal of transition metals from acidic mine waters by novel consortia of acidophilic sulfidogenic bacteria. Microbial Biotechnology. 2012;**5**(1): 34-44

[73] Paschke SS, Banta ER, Dupree JA, Capesius JP. Introduction, conceptual model, hydrogeologic framework, and predevelopment groundwater availability of the Denver Basin aquifer system, Colorado. US Geological Survey Professional Paper. 2011;1-93

[74] Peppas A, Komnitsas K, Halikia I. Use of organic covers for acid mine drainage control. Minerals Engineering. 2000;**13**(5)563-574

[75] Pulford ID. A review of methods to control acid generation of pyritic coal mine waste. In: Davies MCR, editor. Land Reclamation: An end to Dereliction? London: Elsevier; 1991. pp. 269-278

[76] Saarinen T, Mohämmadighävam S, Marttila H, Klove B. Impact of peatland forestry on runnof water quality in areas with sulphide- bering sediments: How to prevent acid surges. Forest Ecology and Management. 2013;**293**:17-28

[77] Sáez-Navarrete C, Rodríguez-Córdova L, Baraza X, Gelmi C, Herrera L. Hydrogen kinetics limitation of an autotrophic sulphate reduction reactor. DYNA. 2012;**172**:126-132

Chapter 9

Fracture Toughness of Concrete Containing Fly Ash

Grzegorz Ludwik Golewski

Additional information is available at the end of the chapter

http://dx.doi.org/10.5772/intechopen.69405

Abstract

The chapter presents results of tests on the effect of the addition of siliceous fly ashes (FA) in the amount of 0, 20, and 30% by weight of cement on the fracture processes in structural concretes. In the course of experiments, measurements of compressive strength of concrete were done as well as fracture toughness for I, II, and III model of cracking was evaluated. During the tests, the effect of age of concretes modified with the additive of fly ashes on analyzed parameters was determined. The experiments were carried out after 3, 7, 28, 90, 180, and 365 days of curing. Fracture toughness of concretes was determined based on the critical stress intensity factors, and then a generalized fracture toughness K_c was determined. The properties of composites with the additive of fly ashes depend on the age of the concrete during tests. Twenty percent additive of fly ashes guarantees high fracture toughness in mature concretes, while the additive of fly ashes in the amount of 30% weight of cement has a beneficial effect on the parameters of concrete only after half a year of curing. Both the 20 and 30% additive of fly ashes significantly reduce the fracture toughness at an early age.

Keywords: fly ash, concrete composite, compressive strength, mode I, II, III fracture, critical stress intensity factor, fracture mechanics, fracture toughness, curing time

1. Introduction

1.1. Purpose and scope of the present study

Among the primary components making up the structure of concrete are Ordinary Portland Cement (OPC), sand, coarse aggregate, water and optional mineral additives, and chemical admixtures. In the last several dozen years, with the development of a new generation of concrete, there has been a significant increase in the production of concrete mixtures containing additives.

A significant increase in the production of concretes with additions is closely connected with the fact that acquisition of natural raw materials and fuels for the economy in the world is becoming an increasing problem. Therefore, a very important issue is to find cheap and fully useful materials of mineral origin that will be able to replace the most expensive component in the composition of cement, which is the Ordinary Portland Cement (OPC). Furthermore, Ordinary Portland Cement is not considered an environmentally friendly material and consumes natural raw materials such as limestone and natural sand. It should be added that the production of Ordinary Portland Cement not only uses a considerable amount of energy, but also emits a substantial amount of carbon dioxide (CO_2) and other greenhouse gases [1]. The production of one ton of Ordinary Portland Cement requires 4 GJ of energy and emits approximately 1.35 billion tons of CO_2 into the atmosphere annually [2]. It is estimated that the CO_2 released during cement clinkering is around 0.7–0.9 tons per ton of Ordinary Portland Cement, meaning that the cement industry generates around 7% of total CO_2 emissions worldwide [3]. For that reason, research on alternative materials has increased. Nowadays, different types of waste are used as additives to concrete in order to improve its durability, strength, and fracture toughness [4]. Using waste materials in concrete has become a necessity to provide a sustainable environment. Composites using waste so-called green concretes [5] or Eco-friendly concretes [4]. Structures made of green concrete are environmentally sustainable and are constructed in such a way that the total impact on the environment during their full life cycle, including service life, is reduced to a minimum.

One of waste materials is fly ash (FA). As a substitute for Ordinary Portland Cement, many types of fly ash are used [6]. However, the most commonly used fly ash is siliceous fly ash, which is a by-product obtained in the process of hard coal combustion performed in electric power stations and in thermal-electric power stations [7]. During combustion processes, siliceous fly ash is removed by mechanical collectors or electrostatic precipitators as a fine particulate residue from the combustion gases before they are discharged to the atmosphere [8].

Nowadays, fly ash has become one of most promising alternative binders to Ordinary Portland Cement. The fly ash from coal combustion processes is a waste very attractive for recycling because concretes with these additives characterized by, higher i.e.:

- corrosion resistance,
- resistance to high temperatures,
- abrasion erosion resistance,

and lower i.e.:

- shrinkage,
- heat of hydration
- water permeability

The siliceous fly ash has been utilized for many years in cement and concrete production. Namely, in the manufacturing process of concretes, it is possible to replace part of the cement binder in the concrete with fly ash treated as pozzolanically active materials. Advantageous impact of fly ash addition is related to the morphology of its particles (shape and surface properties), pozzolanic activity, and microfiller effect (aggregate void filling).

From above reasons, fly ash utilizations are 39% in the USA and 47% in Europe, whereas the global average is estimated to be close to 25%, which is approximately 200 million tons per annum. Up to date, more than 750 million tons of the fly ash are generated each year in the world [9]. The main areas of development of siliceous fly ash are construction, mining, and terrain management.

The analyses of use of this material in the construction industry reveal that more than 50% of siliceous fly ash is used in the cement industry. Therefore, the fly ash is a subject of intensive research. The first paper, which described the properties of concrete containing the fly ash, was published already in (1937) [10], whereas the first critical review of using fly ash in concrete was published in 1980 [11].

Nowadays, good-quality siliceous fly ash is used for production of Portland-composite cement and plain concretes as well as self-compacting concretes, roller-compacted concretes, or steel-fiber reinforced concretes. Moreover, with the right quality (low loss of ignition (LOI) and high fineness) and quantity (up to 15%) of the fly ash, selection is also possible to use them for the production of polymer concrete and concretes with higher strengths, that is, high-performance concretes and ultra-high-performance concretes.

What effect does the addition of siliceous fly ash have on the basic properties of the concrete mixture and concrete, and how it should be used are described in detail in several monographs [12–14] and numerous scientific publications. Nevertheless, certain features of concrete composites with fly ash have so far been investigated and analyzed very rarely. For example, up till now the effect of fly ash influence on the fracture toughness of the concrete was not investigated systematically. Such studies were carried out rather occasionally.

Therefore, this chapter presents results of macroscopic tests on the effect of the addition of siliceous fly ashes in the amount of 0, 20, and 30% by weight of cement on the fracture processes in structural concretes. In the course of experiments, measurements of compressive strength were done as well as fracture toughness for I, II, and III model of cracking was evaluated. Fracture toughness of concretes was determined based on the critical stress intensity factors for all mode fracture, and then a generalized fracture toughness K_c was determined. During the tests, the effect of age of concretes modified with the additive of siliceous fly ashes on analyzed mechanical parameters was determined. The experiments were carried out after 3, 7, 28, 90, 180, and 365 days of curing.

1.2. Discussion of topic selection and scientific objectives

The properties of concrete materials, including their durability, mainly depend on the structural factors and the interaction between micro- and macro-structure of the material. Cracks and losses are the two basic defects of the concrete structure, which may reduce the load-bearing capacity, leak tight integrity, and stiffness of elements and structures, which in extreme cases may lead to failure and even building catastrophes. It is therefore important for the engineer designing the concrete mix to be familiar with the processes of damage and crack development in structural materials, especially those that are characterized by high brittleness. It allows improvement of the quality of concrete, from which the structures are made, estimation of defects, and determination of their causes. This leads to obtaining composites with the highest durability and reliability.

The overall objective of the conducted scientific analyses was a description of fracture processes in concretes with the siliceous fly ash additive. Concretes with these additives have been used in the industry for approximately 80 years, and in the past two decades, their use in the composites with cement matrix has significantly increased. Therefore, it is obvious that in the course of such a long period of time, it has managed to explore the effect on the basic physical and mechanical characteristics of the siliceous fly ash additive. However, based on literature research, it is clear that in this area of science, there is a clear gap. It is linked to the lack of complete data relating to the analysis of initiation and propagation of microcracks and cracks, in this type of materials, developing during the impact of complex states of stress on concrete.

Description of gradual degradation of brittle materials like concretes is a multi-stage process depending on an internal structure of composite and a level of external loads. A detailed description of particular stages of development of cracks in concrete structure is shown in Section 2. The reduction of strength of concrete results from initial structural defects existing inside material [15]. Additionally, a crucial role on the fracture behavior of concrete plays an aggregate structure [16].

Cracks initiation and propagation in the concrete require the knowledge of all fracture mechanics parameters for modes I, II, and III, that is, K_{Ic}, K_{IIc}, and $KIII_c$. This is due to the fact that fracture is an important feature in concrete at all scales [17]. Moreover, the damage of concrete subjected to complex loading involves strong anisotropy due to its highly heterogeneous nature and geometrically anisotropic characteristic of microcracks. Thus, the three-dimensional (3D) fracture process is generally complicated, that is, an experimental estimation of all fracture mechanics parameters is extremely difficult and can be done for three separate fracture modes: I, II, and III. **Figure 1** defines the three modes of loading, that is:

- Mode I, opening or tensile mode,
- Mode II, sliding or in-plane shear (pure shear) mode, and
- Mode III, tearing or anti-plane shear mode.

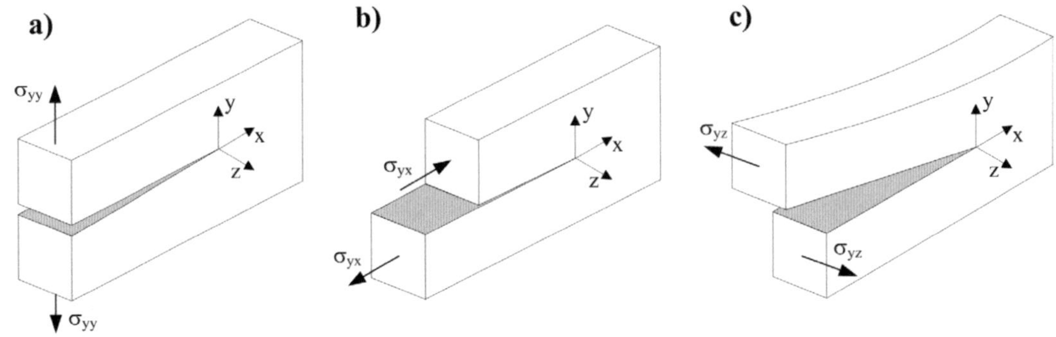

Figure 1. Schemes of the three modes of cracking: (a) mode I, (b) mode II, and (c) mode III.

In general, cracks in real applications are often subjected to a mixed mode loading conditions, whereas fracture may occur under mode I, mode II, mode III, or a combination of them [18]. The result is that in many cases, cracks in structural elements start and develop in the mixed mode I–II loading or I–III loading. A detailed analysis of stress intensity factors for plain concrete is presented by Song et al. [19] for different combinations of pure and mixed modes of cracking to formulate general three-dimensional (3D) fracture criterion. Tests of fracture toughness in complex stress states are associated with the description of the present critical state in the element, which can be written as the following relationship:

$$f(K_{Ic}, K_{IIc}, K_{IIIc}) = f_c. \quad (1)$$

With the known values of K_{Ic}, K_{IIc}, and K_{IIIc}, it is possible to designate a generalized fracture toughness of the material, K_c, from Eq. (2) [20]. The reason that the stress intensity factor is useful is because fracture (sudden crack growth to failure) occurs when K exceeds a certain critical value, K_c, called exactly the fracture toughness.

$$K_c^2 = K_{Ic}^2 + K_{IIc}^2 + \frac{K_{IIIc}^2}{1 - \nu}, \quad (2)$$

where ν is the Poisson's ratio.

Unfortunately, in the assessment of fracture toughness of concretes with fly ash, the available results from experiments relate only to the first mode fracture, that is, tension at bending. Fracture toughness under mode I for plain and high-performance concretes containing siliceous fly ash additives, applying three-point-bending tests, was reported in Refs. [21–23]. No article was found in the literature that would discuss the fracture toughness of concrete with the siliceous fly ash additive under mode II and III loading.

Lack of comprehensive data describing the effect of siliceous fly ash additive on the fracture processes in structural concretes caused that **the main reason** to raise the subject of scientific work was experimental determination of fracture toughness of concrete—in which a portion of the binder was replaced with siliceous fly ash—under all three modes fracture.

The second reason is that encouraging the commencement of work on the analysis of the fracture processes in concretes with the siliceous fly ash additive was closely associated with the phenomenon of the variable dynamics of increasing the strength of such materials in the process of curing [24]. Due to the slow course of the pozzolanic reaction that has a direct influence on the mechanical properties of composites of this type, the increase in the strength at the initial stage of hardening is slow [25]. However, in the longer time of curing, the durability of cement with the siliceous fly ash reaches values exceeding the compressive strength of Portland Cement of the same strength class. In relation to the second reason of working on this particular subject of the scientific paper, efforts were made to analyze the concrete at early age and in a period exceeding 28 days:

- changes in compressive strength of concrete and
- changes in parameters of fracture mechanics of concretes.

All experiments were planned for two compositions of concrete mixture, with varying percentage of siliceous fly ash additive, which is often used in the cement industry. An assumption in the analysis in the target range resulted from the following fact. In tests, of which the results were presented in paper [8], fracture toughness under II mode fracture was analyzed in concretes with the siliceous fly ash additive in the amount of 10, 20, and 30% of weight of cement. On their basis, it was found that 10% siliceous fly ash additive has a minor impact on the value of parameter K_{IIc} causing it to increase by only 1%, while 20 and 30% additive of this waste significantly changes the fracture toughness. Therefore, all tests were conducted for concretes modified with the siliceous fly ash additive in the amount of 20 and 30% of weight of cement. The obtained results from experiments were compared to the values obtained for the reference concrete, which was a composite made with the use of cement CEM I.

2. Scope of the studies and methodology of the tests

2.1. Scope of the studies

In order to accurately understand the fracture processes in the concretes with the siliceous fly ash additive, an extensive program of laboratory tests on a wide range of diagnostic methods was developed. In the first stage, properties of siliceous fly ash used in concretes were characterized. In testing, an attention was paid to parameters of fly ash such as:

- chemical and phase composition,
- bulk density,
- specific surface area,
- pozzolanic activity, and
- natural radioactivity.

Then, for all planned testing of concrete, specimens with the siliceous fly ash additive were prepared. The fracture toughness tests to estimate critical stress intensity factors and the compressive strength tests were conducted with three types of concretes of different volume content of the FA additive (like in Refs. [8, 26–29]), that is:

- without the FA additive (FA-00),
- with 20% FA additive (FA-20), and
- with 30% FA additive (FA-30).

Concrete testing included:

- analysis of compressive strength parameter and
- analysis of parameters of fracture mechanics.

In the course of the work, macroscopic measurements of compressive strength and fracture toughness for all modes were conducted. A w/c ratio was constant in all tested concretes. In modified concretes, cement was replaced by fly ash by its weight. For each of the three composites, all experimental tests were conducted in six time periods, with the age of concrete being 3, 7, 28, 90, 180, and 365 days.

Advanced diagnostic equipment was used during the experiments. For the evaluation of fracture toughness, three separate test stands were organized (see Section 5). On the basis of the planned scope of works, associated with the realization of the academic achievement, complete repository of **detailed academic objectives** is as follows:

a. In the range of fly ash testing, the scientific objectives were:

- analysis of physicochemical properties of fly ash and
- evaluation of pozzolanic activity of fly ash.

b. In the range of fracture toughness, the scientific objectives were:

- experimental determination of the impact of siliceous fly ash additive on the critical stress intensity factor of concrete under I, II, and III mode fracture, between the 3rd and 365th days of curing and
- determination of particular concretes with generalized fracture toughness K_c.

Due to the transparency of this chapter, this section is limited only to provide general information related to the experiments. Detailed descriptions of the strength tests were presented in Section 4, while fracture toughness tests are presented in Section 5. The basic properties of used fly ash are described in section 3.

2.2. Materials used in the studies and methods used to investigate them

2.2.1. Materials used in the preparation of mixtures

In elaboration of a new concrete mixture having the fly ash additive, it was assumed that the total amount of binding material in the concrete should be constant. Therefore in this methodology, a certain portion of cement is replaced in the mixture by the same amount of the fly ash. The following materials were used for making the mixtures:

- Ordinary Portland Cement (OPC) CEM I 32.5 R,
- natural gravel aggregates of maximum grain size up to 8 mm,
- a pit sand with a maximum diameter of 2 mm,
- siliceous, class F, fly ash from local power plant is a result of energetic combustion of hard coal in the Puławy thermal-electric power station, and
- Basf Liquol BV-18 plasticizer (based on calcium lignosulphonates); 0.6% of binding material weight.

2.2.2. Binders investigations

The chemical compositions of the Ordinary Portland Cement as well as the fly ash were evaluated by X-ray fluorescence (XRF), whereas microstructure of these materials was evaluated by using Scanning Electron Microscope (SEM). The microstructural testing was carried out using a QUANTA FEG 250 at magnification from 200 to 80,000 times equipped with an energy dispersive Spectroscopy (EDS EDAX). The tests were performed in both the low and high vacuum. A bulk density in both materials was determined by the pycnometric method, whereas the specific surface area was determined by the Blaine method. The average value of the two tests was determined based on the three results.

2.2.3. Properties of the materials

The main parameters of the materials used in the studies are given below.

2.2.3.1. Ordinary Portland Cement

The physical parameters of Ordinary Portland Cement are collected in **Table 1**, whereas the chemical constituents are shown in **Table 2**.

The mineralogical compositions of Ordinary Portland Cement were analyzed by the Bogue method. They are listed in **Table 3**.

2.2.3.2. Aggregates

The basic properties of aggregates are shown in **Table 4**, whereas the particle size distribution in **Table 5**.

An attempt was made to select the optimum proportion of different-sized aggregates in such a way as to be contained in the most advantageous area between limiting grain size distribution curves. The recommendations used were based on the German standard DIN 4226-1.

2.2.3.3. Admixture

In order to improve the workability of concrete mixtures, plastifying admixture composed on the basis of calcium lignosulfonate with a density of 1.16 g/cm^3 and the dosing range of

Physical parameters of cement		Average values
Specific surface area according to Blaine (cm^2/g)		3280
Bulk density (g/cm^3)		3.11
Initial setting time (min)		207
Final setting time (min)		298
Compressive strength (MPa)	After 2 days	23.3
	After 28 days	50.0
Volume change (mm)		0.56

Table 1. Physical parameters of Ordinary Portland Cement CEM I 32.5 R.

Constituent	Contents (% mas.)
SiO_2	21.37
Al_2O_3	5.02
Fe_2O_3	2.40
CaO	63.95
MgO	2.47
SO_3	3.0
Na_2O	0.18
K_2O	0.91
Cl	0.057
Insoluble residues	1.11
Loss of ignition	1.24

Table 2. Chemical composition of the Ordinary Portland Cement CEM I 32.5 R.

C_3S	C_2S	C_3A	C_4AF	$CaSO_4$ (gypsum)
60.69	15.82	9.24	7.28	5.10

Table 3. Mineral components of the Ordinary Portland Cement CEM I 32.5 R (%).

Property	Unit	Aggregate type	
		Fine aggregate—sand	Coarse aggregate—gravel
Specific density	g/cm^3	2.60	2.65
Bulk density	g/cm^3	2.20	2.25
Compressive strength	MPa	33	34
Modulus of elasticity	10^2 MPa	330	330
Sand point for mix	%	40.7	

Table 4. The basic properties of aggregates used in the study.

0.1–1.0% of mass of cement was used. The plasticizer is used in an amount of 0.6% of mass of the binder.

2.3. Mixing casting and specimens curing

A mix proportion of the concrete compositions are presented in **Table 6**. All mixtures had the same water-binding material indicator $w/b = 0.4$. The cast specimens were covered with a polyurethane sheet and damped cloth. They were placed in 20 ± 2°C chamber. After 2 days, all

Fraction (mm)	Content of aggregates fraction (%)		
	Sand	Coarse aggregate	Mix
0–0.125	2.9	0.7	1.7
0.125–0.25	14.8	0.4	5.6
0.25–0.5	41.1	0.4	15.3
0.5–1.0	32.7	1.6	12.4
1.0–2.0	4.5	6.9	5.7
2.0–4.0	4.0	19.9	13.9
4.0–8.0	0.0	63.1	40.2
8.0–16.0	0.0	7.0	5.2
Sand point	96.0	10.0	40.7

Table 5. The particle size distribution of the aggregates used.

Concrete	Cement (kg/m^3)	Fly ash (kg/m^3)	Water (kg/m^3)	Sand (kg/m^3)	Coarse aggregate (kg/m^3)	Plasticizer (kg/m^3)
FA-00	352	0	141	676	1205	2
FA-20	282	70	141	676	1205	2
FA-30	246	106	141	676	1205	2

Table 6. Mix proportion of the concrete compositions.

specimens were demolded. Then, they were kept for the first 14 days in a chamber with a moisture-saturated atmosphere. During the next days to the study after 28, 90, 180, and 365 days, specimens were cured in laboratory conditions (20 ± 2°C). In case of specimens tested after 3 and 7 days, they were removed from the water at least a few hours before the study. After a suitable period of curing, the compressive strength tests and other basic tests were carried out.

2.4. Specimens used in the studies

When deciding about the number of specimens for experiments, attempts were made to reach a compromise between the cost of their preparation as well as the minimum amount necessary to guarantee reliable statistical values and the ability to generalize the results obtained from specimens on the entire population of concrete. Therefore, in all macroscopic tests, average values of the conducted experiments were assessed on the basis of results obtained for six specimens. Assortment of specimens for compressive strength and fracture toughness tests of concrete, for each of the mixtures in all time periods, was as follows:

- 6 cubic specimens (150 mm) for testing the compressive strength—f_{cm},
- 6 beams (80 × 150 × 700 mm) with one initial crack for testing fracture toughness under mode I fracture—K_{Ic}^{S},

- 6 cubic specimens (150 mm) with two initial cracks for testing fracture toughness under mode II fracture—K_{IIc}, and
- 6 cylindrical specimens with a diameter of 150 mm and a height of 300 mm having an initial circumferential notch, for testing fracture toughness under mode III fracture—K_{IIIc}.

2.5. The concept of analysis of results obtained in experimental tests

When compiling the results of strength and fracture toughness tests, tables for further analysis for each type of the conducted experiments included average values and standard deviations—δ. Furthermore, graphs presenting the summary of the given parameter for all series of concrete, in all time periods, were prepared for average values in particular tests. The relationships of the analyzed indicators as a function of the age of concrete were shown graphically in order to better illustrate the changes occurring in concretes during their curing. Percentage changes of parameters, during curing of composites, were also shown, taking the value obtained after 365 days as 100%.

Results of the fracture toughness, for all modes of cracking, and generalized fracture toughness are summarized in **Table 11**.

3. Characteristics of used siliceous fly ash

3.1. Microstructure of cementitious materials

Figure 2 shows the morphologies and the particle sizes of the cementitious materials (cement and fly ash) by SEM observations. In order to accurately compare the microstructure of both fillers in the SEM studies, the same scale and magnifications were used (**Figure 2**).

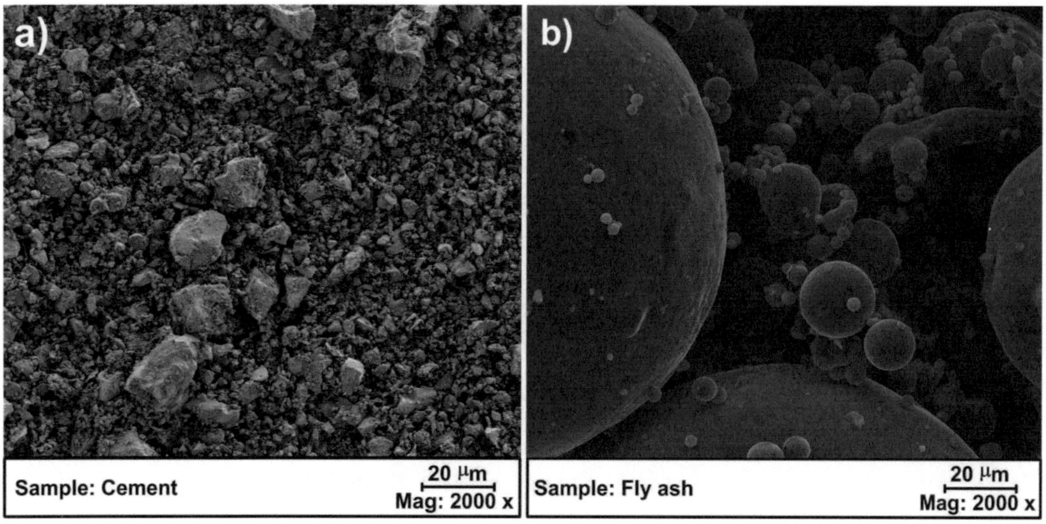

Figure 2. SEM images of cementitious materials: (a) cement and (b) fly ash.

Other interesting characteristic pictures of a typical grains and microstructures of the used fly ash at various magnifications were shown in paper [8]. On the basis of analysis of SEM images (**Figure 2(b)**), it can be concluded that fly ashes are characterized by oval regular shapes of fine particle size.

3.2. Chemical composition of the fly ash

The chemical constituents of the fly ash are shown in **Table 7**. The fly ash is a class F with 85.09% of $SiO_2 + Al_2O_3 + Fe_2O_3$, 0.65% of SO_3, and 3.2% of loss of ignition (LOI) meeting the requirement of ASTM C618.

Constituent (wt. %)	SiO_2	Al_2O_3	CaO	MgO	SO_3	Fe_2O_3	LOI
Fly ash	50.96	25.88	2.15	2.6	0.65	8.25	3.2

Table 7. Chemical composition of the fly ash.

Parameter	Average values
Specific surface area according to Blaine (cm^2/g)	3640
Bulk density (g/cm^3)	2.14

Table 8. Physical parameters of the fly ash.

3.3. Physical parameters of the fly ash

The main physical parameters of the fly ash are shown in **Table 8**.

Physical properties of the cement and the fly ash, and **Figure 2** suggest that the fly ash is finest, followed by the cement. Small particle size of the fly ash has a beneficial effect on compressive strength, durability, and permeability of concretes with these additives [30].

3.4. Mineral composition of the fly ash

The crystalline phases of the fly ash were identified by X-ray diffraction (XRD) analysis [8]. The crystalline phases of the fly ash were identified by XRD patterns. Randomly oriented powder specimen (about 1 g weight) for XRD analysis was prepared. XRD graphs were obtained at room temperature by a PANalytical X'Pert PRO MPD diffractometer (with the PW 3050 goniometer), Cu lamp, and graphite monochromator. The diffractogram from 5° to 65° in the $2\theta$and scanning rate was at 0.05° intervals. PANalytical X'Pert HighScore and the ClayLab ver. 1.0 software were used to process diffraction data. The identification of mineral phases was based on the PCPPDFWIN ver. 2.1 database formalized by the JCPDS-ICDD.

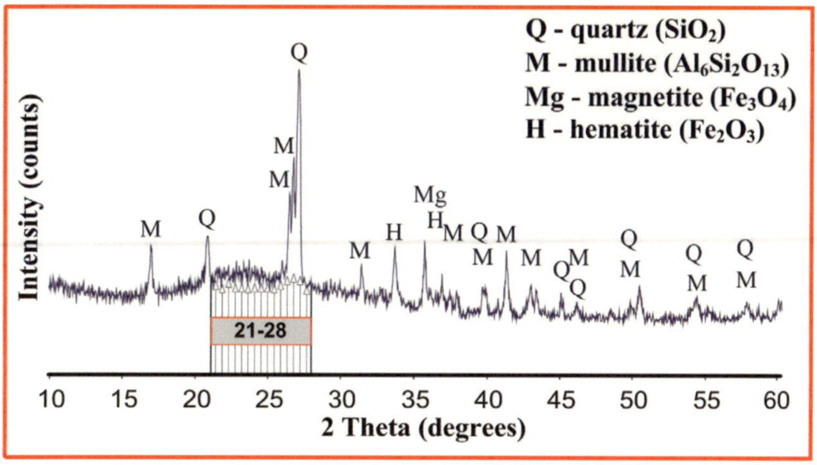

Figure 3. XRD pattern of the fly ash.

The X-ray diffractogram of the fly ash (**Figure 3**) shows that there are, beside glass, four major crystalline components in the phase composition of the fly ash: quartz (SiO_2), mullite ($Al_6Si_2O_{13}$), magnetite (Fe_3O_4), and hematite (Fe_2O_3). On the diffractogram, a characteristic increase in intensity at the angle ranging from 21 to 28° 2θ can be observed as well as relatively low intensity of the peaks of quartz and mullite phases. This image of the phase composition of the fly ash indicates their fine particle size and a high-strength activity index.

3.5. Pozzolanic activity of the fly ash

The applicability of the fly ash depends mainly upon its pozzolanic activity. In order to assess the pozzolanic activity of the used FA were selected physical method, according to EN 4501-1 standard. EN 450-1 defines as a standard pozzolanic activity index of the fly ash the strength activity index which is the ratio (in %) of the compressive strength of mortar containing 75 wt% of cement CEM I 42.5R, 25 wt% of the fly ash and cement mortar without addition. According to EN 450-1 standard, the pozzolanic activity index of the fly ash is determined after 28 and 90 days of hydration. The compressive strength of cement is measured according to procedure described in EN 196-1 standard, using 40 × 40 × 160 mm prisms of the mortar. The water-to-solid ratio in mortars is 0.5. According to the EN 450-1 standard, the acceptable level of pozzolanic activity is attained when the 28-day compressive strength of the fly ash and cement mortar constitutes 75% of the value for the control sample and after 90 days—85%, respectively.

Because compressive strength tests and fracture toughness test were conducted in six time periods in the range between 3 and 365 days, pozzolanic activity index was also determined at the same time intervals, that is, after 3, 7, 28, 90, 180, and 365 days. **Table 9** shows the average values of the strength activity index, whereas **Figure 4** shows how it looked its growth rate.

Age (days)	Pozzolanic activity index (%)
3	56.73
7	68.36
28	92.13
90	111.84
180	116.72
365	123.53

Table 9. The assessment of pozzolanic activity by EN-450-1 standard.

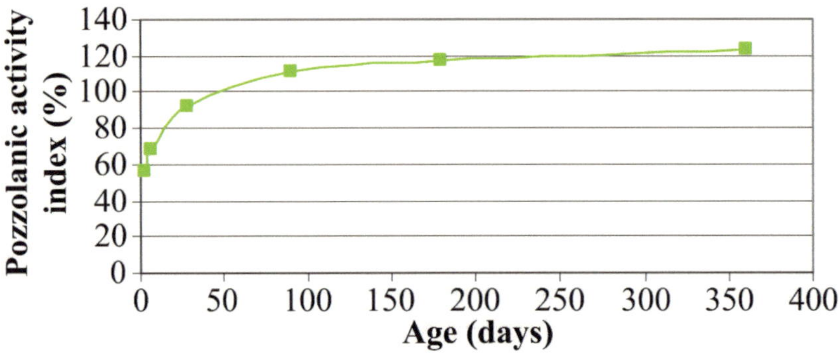

Figure 4. Pozzolanic activity index vs. age of mortar.

A line connecting points on the graph (**Figure 4**) shows a gradual, non-linear increase of the activity of the fly ash between 3rd and 365th days of curing of samples. After 3 days, the strength activity index is very low, which is certainly due to the slow growth of the reaction products after such a short period of curing. According to Lee et al. [31], siliceous fly ash after 1 day is still inactive, and according to Ogawa et al. [32], pozzolanic reaction in this material begins in 1–3 days. Also Fraay et al. [33] note that up to 1 week, the fly ash does not show a distinct amount of pozzolanic reactivity.

Based on the strength activity index (**Table 9**), it can be concluded that the initial effect of the pozzolanic reaction (consisting in strengthening the matrix) in the composites, which include the addition of siliceous fly ash, appears until after 7 days of curing. This is also confirmed by other studies [34]. After this time, it can be seen a significant increase in the activity of the material—more than 20%. In the period between the 7th and the 28th days is already visible intense increase in the strength of mortars with fly ash and the activity index compared to the value of weekly increases by almost 35%. A significant increase in the analyzed parameter is visible even between the 28th and the 90th days of curing of samples. In a further period of time, between 3 and 6 months and after a year, there has been a slight increase of the pozzolanic activity of fly ashes (by a few percent). It can be concluded that after 6 months, the ongoing reactions stabilizes. Nevertheless, the reactions still exist even after 365 days [34].

As a result of the study, it can be concluded that the pozzolanic activity of the fly ash meets the requirements of EN 450-1 standard. Strength activity index was 92.13% at 28 days and 111.84% after 90 days, which means that it exceeds significantly the minimum values specified in the requirements of the EN-450-1 standard.

3.6. Natural radioactivity of the fly ash

Natural radioactivity of the fly ash and concrete made with their addition was analyzed based on concentration of radioactive elements (potassium, radium, and thorium), radioactivity coefficients f_1 and f_2, and the radiation dose [35]. The study of natural radioactivity of used fly ash shows that these materials are not hazardous from the radiological point of view. Both concentration of radioactive elements and radioactivity coefficient and radiation dose were contained within acceptable limits [35].

4. Compressive strength analysis

The uniaxial compression strengths were tested using a compression machine (Walter + Baiag) with a maximum load of 3000 kN. The loading rate of compressive strength test was controlled between 0.5 and 0.8 MPa/s. The compressive strengths f_{cm}, were tested with application of cubic specimens (described in Section 2.4) according to the standards of series EN 12390.

Table 10 summarizes the results from compressive strength tests of concretes, for particular periods of curing, while **Figure 5** presents relationships f_{cm}, as a function of the age of concrete.

Fly ash additive caused a clear decrease of compressive strength in concretes analyzed at an early age, that is, after 3 and 7 days. According to **Table 10**, value f_{cm} after 72 h of curing was almost 8 and exactly 10 MPa higher in concrete without the fly ash additive in comparison to FA-20 and FA-30, respectively. After a week, the differences between the reference and modified concretes were only 3 MPa. According to Ref. [26], in which a more detailed analysis

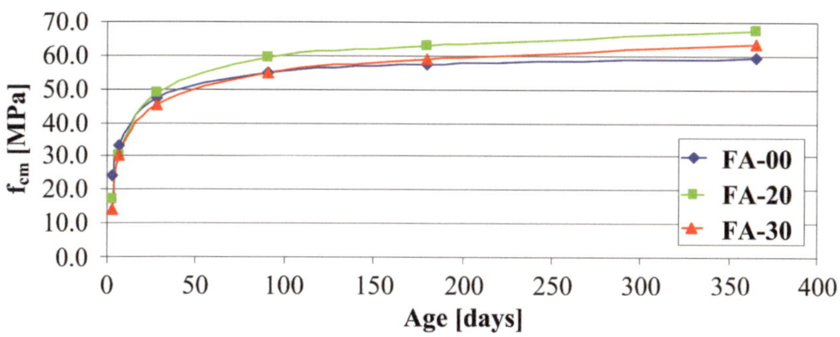

Figure 5. Compressive strength of analyzed concretes as a function of curing time.

Concrete	Age (days)	$f_{cm} \pm \delta$ (MPa)
FA-00	3	24.23 ± 2.60
	7	33.18 ± 2.57
	28	47.51 ± 2.55
	90	55.13 ± 2.51
	180	57.22 ± 2.48
	365	59.25 ± 2.46
FA-20	3	16.95 ± 3.05
	7	30.12 ± 3.03
	28	48.96 ± 3.02
	90	59.35 ± 2.80
	180	62.81 ± 2.52
	365	67.29 ± 2.35
FA-30	3	14.23 ± 3.59
	7	30.06 ± 3.57
	28	45.10± 3.55
	90	55.11 ± 3.10
	180	58.83 ± 2.86
	365	63.27 ± 2.50

Table 10. Compressive strength of concretes.

of changes of strength parameters in concretes with the fly ash additive at an early age is presented, clear disproportions in the obtained results occur within 14 days of curing. After 3 weeks, 20% fly ash additive strengthens the structure of concrete composites to the extent that the values of f_{cm} are slightly higher in this type of concrete in comparison to FA-00. Also, after 28 days and in subsequent time periods, FA-20 had by far the highest strength, which was probably due to the rapid increase of pozzolanic reaction products after a longer time of curing. Concrete with a larger amount of fly ashes was characterized by the lowest strength in the period up to 3 months. After a half year of curing, value of f_{cm} for this composite was higher in comparison to concrete without the additives; however, it still lower in comparison to FA-20. A further increase of strength of FA-30 caused that after a year strength of this material was 4 MPa higher f_{cm} in comparison to FA-00 and lower by the same value in comparison to FA-20.

The conducted own tests have shown that compressive strength of concrete was increasing with time. The growth dynamics of this parameter, however, differed significantly in particular types of analyzed composites. This can be easily observed by comparing relative changes of compressive strength over time, which is shown in **Figure 6**.

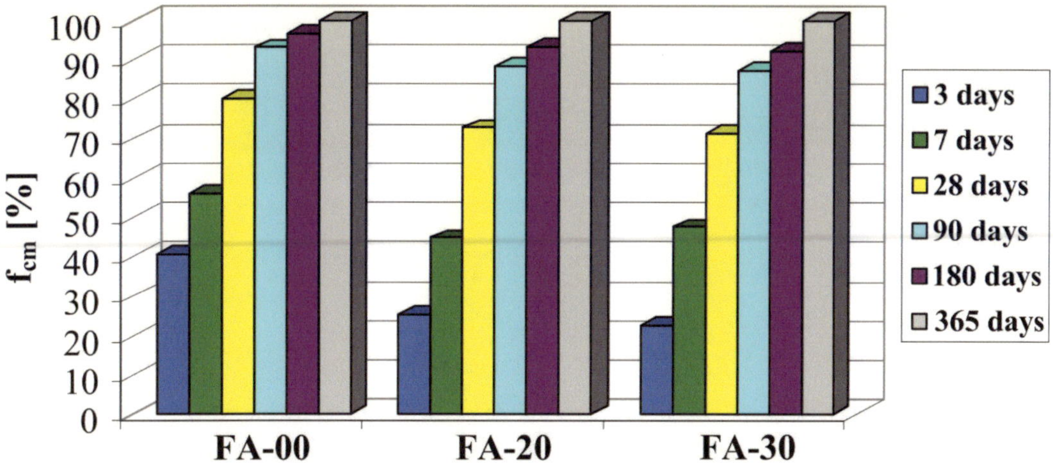

Figure 6. Relative changes of compressive strength over time.

After the first measurement (3 days), concrete FA-00 had more than 40% of the annual compressive strength, whereas in concretes with fly ash additives, this strength did not even had 30% of the final strength. After a week, concrete, without the fly ash additive, was characterized by a higher relative strength, however, in the analysis of obtained values, a clearer dynamics of strength increase in composites with fly ashes was observed. Although reference concrete had already more than 50% of its 365-day strength, which was a result 10% better in relation to concretes with additives, a greater increase of f_{cm} was observed in concretes with fly ashes between 3rd and 7th days of curing. Also, between 7th and 28th days of curing, concretes with additives are characterized by significant increments of f_{cm} (amounting to: FA-20—50%, FA-30—62%), which correlates with the activity of pozzolanic fly ashes obtained with the use of physical method in accordance with EN 450-1 (**Table 9**, **Figure 4**). Clear growth trends of compressive strength, in modified composites, were observed even after 90 days after the start of tests. Both 20 and 30% fly ash additives caused increase of f_{cm} within 2 months by more than 20%, which was a result 6% better in relation to FA-00. After half a year in all types of composites, the increase of compressive strength was small and was 6% for modified concretes and only 3% in the reference concrete. A similar trend was also observed in relation to 1-year concretes where the strength increase in concretes with fly ashes was 7% and was higher by half in comparison to the results obtained for FA-00.

5. Fracture toughness tests

5.1. Fracture toughness tests during mode I loading

The testing of Mode I fracture toughness was performed according with the draft guidelines of International union of laboratories and experts in construction materials, systems and structures (RILEM) recommendations [36]. The fracture toughness of the composites with the fly ash addition was determined based on the experimental results of critical stress intensity factor—K_{Ic}^{S}.

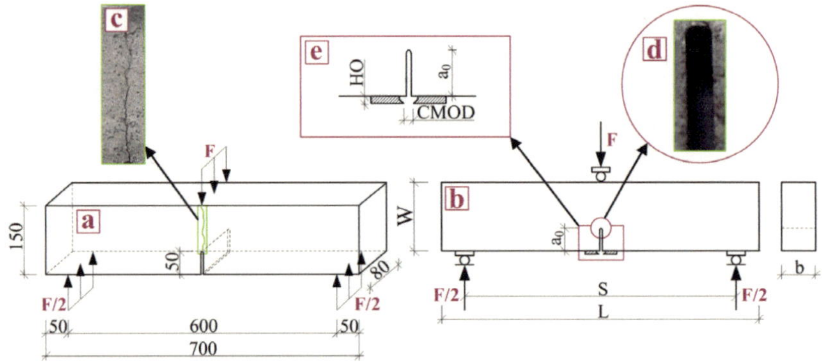

Figure 7. Specimen with (a) static scheme, (b) geometry, (c) exemplary crack paths, (d) blunted notch, and (e) details of calliper gauge holder.

To assess the fracture toughness of concrete, beams with the dimensions 700 × 150 × 80 mm (**Figure 7**) which had one initial centrally crack were used (see Section 2.4). The beams were subjected to three-point bending test. They were made in demountable bolted wooden forms. The assumed size of the initial crack in the beams was achieved by actually concreting flat steel plates, having a thickness of 3 mm. A special experimental stand was prepared as shown in [28]. All necessary results, needed to determine the critical stress intensity factors for concretes, were obtained with application MTS 810 testing machine. The width of the initial crack opening during the tests was measured using a crack opening sensor, that is, the MTS clip gage axial extensometer, which was placed on the clamping test grips (**Figure 7**).

The specimens placed in the experimental stand were subjected to cyclic loading process performed quasi statically. The maximum force (95% of the fracture load) in the one cycle was achieved in 5 min. Then, we start unloading process up to 0 kN. After that the cycles were repeated until the beams were broken into two parts. The whole cyclic deformation processes were described by the following curves:

- Force (*F*)—crack mouth opening displacement (*CMOD*) and
- Force (*F*)—deflection (*f*).

The obtained results (which are summarized in **Table 11**) allowed for determination of the fracture toughness K_{Ic}^{S} using the formula given in RILEM recommendations [35].

5.2. Fracture toughness tests during mode II loading

In order to assess the influence of the fly ash volume content on the shear fracture toughness in the considered concrete composites, compact shear specimens (CSS) were prepared for the basic tests [8, 16, 26, 27]. Therefore, cubes with 150 mm long edge and two fictitious cracks were used for the experiments (see Section 2.4). The target crack sizes were obtained by embedding in concrete the cubes being formed by two 4 mm steel sharpened flat plates. The experiments were performed for the loading scheme presented in **Figure 8**. Specimen

Concrete	Age (days)	$K^S_{Ic} \pm \delta$ (MN/m$^{3/2}$)	$K_{IIc} \pm \delta$ (MN/m$^{3/2}$)	$K_{IIIc} \pm \delta$ (MN/m$^{3/2}$)	K_c(MN/m$^{3/2}$)
FA-00	3	0.58 ± 0.08	2.26 ± 0.38	1.36 ± 0.19	2.70
	7	0.79 ± 0.09	3.18 ± 0.41	1.88 ± 0.18	3.78
	28	1.06 ± 0.10	4.24 ± 0.40	2.49 ± 0.20	5.03
	90	1.21 ± 0.08	4.93 ± 0.35	2.92 ± 0.19	5.86
	180	1.26 ± 0.06	5.12 ± 0.30	3.06 ± 0.17	6.10
	365	1.30 ± 0.05	5.31 ± 0.24	3.19 ± 0.14	6.33
FA-20	3	0.40 ± 0.09	1.48 ± 0.34	0.92 ± 0.19	1.79
	7	0.73 ± 0.09	2.93 ± 0.40	1.75 ± 0.21	3.49
	28	1.09 ± 0.11	4.39 ± 0.51	2.60 ± 0.24	5.22
	90	1.31 ± 0.10	5.33 ± 0.41	3.16 ± 0.22	6.33
	180	1.39 ± 0.07	5.70 ± 0.32	3.36 ± 0.19	6.76
	365	1.48 ± 0.06	6.14 ± 0.29	3.59 ± 0.18	7.26
FA-30	3	0.30 ± 0.08	1.05 ± 0.25	0.68 ± 0.16	1.29
	7	0.62 ± 0.08	2.40 ± 0.31	1.41 ± 0.19	2.85
	28	0.93 ± 0.10	3,65 ± 0.42	2.14 ± 0.21	4.33
	90	1.17 ± 0.09	4.68 ± 0.38	2.77 ± 0.20	5.56
	180	1.28 ± 0.08	5.16 ± 0.34	3.08 ± 0.18	6.14
	365	1.37 ± 0.07	5.58 ± 0.27	3.31 ± 0.17	6.63

Table 11. Fracture mechanics parameters of concretes.

dimensions in detail and exemplary crack paths, which were observed during experiment, were also shown in **Figure 8**.

The tests were performed on the MTS 810 servo-hydraulic testing machine, like in the studies during mode I loading. Fracture toughness of the concretes was determined on the basis of analysis of critical stress intensity factor K_{IIc}. This factor was determined according to the formula proposed by Watkins [37], whereas average test results are summarized in **Table 11**.

5.3. Fracture toughness tests during mode III loading

In order to determine the fracture toughness K_{IIIc}, cylindrical specimens with a diameter of 150 mm and a height of 300 mm with an initial circumferential notch of 2 mm thickness (made in the half-height) were tested. The depth of the notch was equal to ¼ of the cylinder diameter. Initial notches in the specimens were created during their formation by application of the two semi-circular steel inserts placed in the half-height of specially prepared cylindrical forms [29].

The experimental stand (**Figure 10**) for the fracture toughness test at the mode III fracture consisted of the cylindrical specimen with the initial notches, steel plates, and screws with washers securing the specimen in the press holders, tension-torsion MTS testing machine [29].

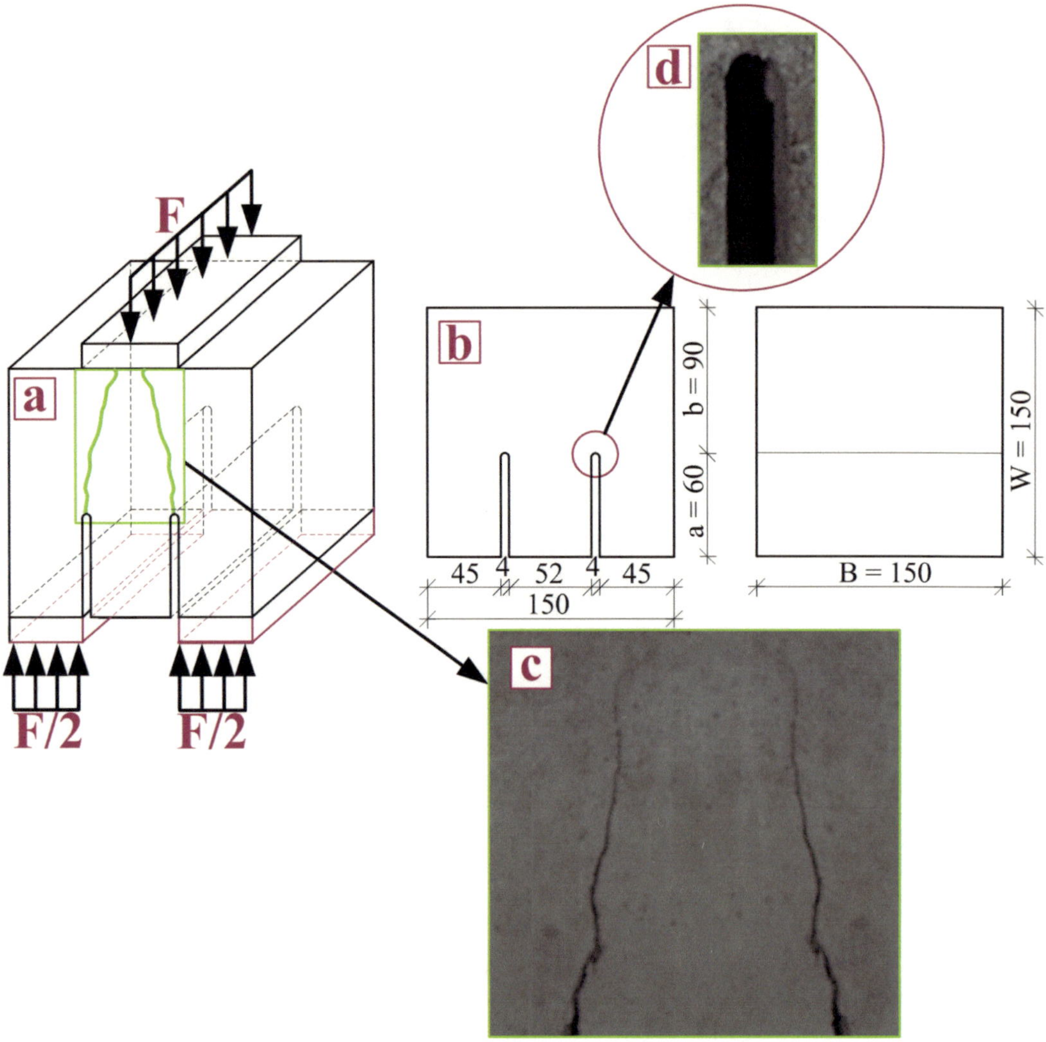

Figure 8. Specimen with (a) loading conditions, (b) dimensions (in mm), (c) exemplary crack path, and (d) blunted notch.

To mount the specimens in the grips of the torsion testing machine, two types of circular steel plates were designed and manufactured. During forming, the bottom plates of 15 mm thickness (**Figure 9**) were anchored in the specimens on their top and bottom with 6 bolts M12/65. Then, top plates of 10 mm thickness (**Figure 9**) were bolted to these plates with 6 bolts M12/20. Top plates hold the specimen directly in the grips of the torsion testing machine from the top and bottom with M28/70 bolts. The full device for testing the K_{IIIc} is shown in **Figure 9**.

Specimens were tested on the axial tension-torsion testing machine MTS 809 in accordance with the load diagram shown in **Figure 10**. The shear process of specimens was controlled by rotation angle ω assuming a small increase equal to 0.5°/min. Critical values of a torque and the rotation angle which corresponds to the failure of the specimens were reached after a few

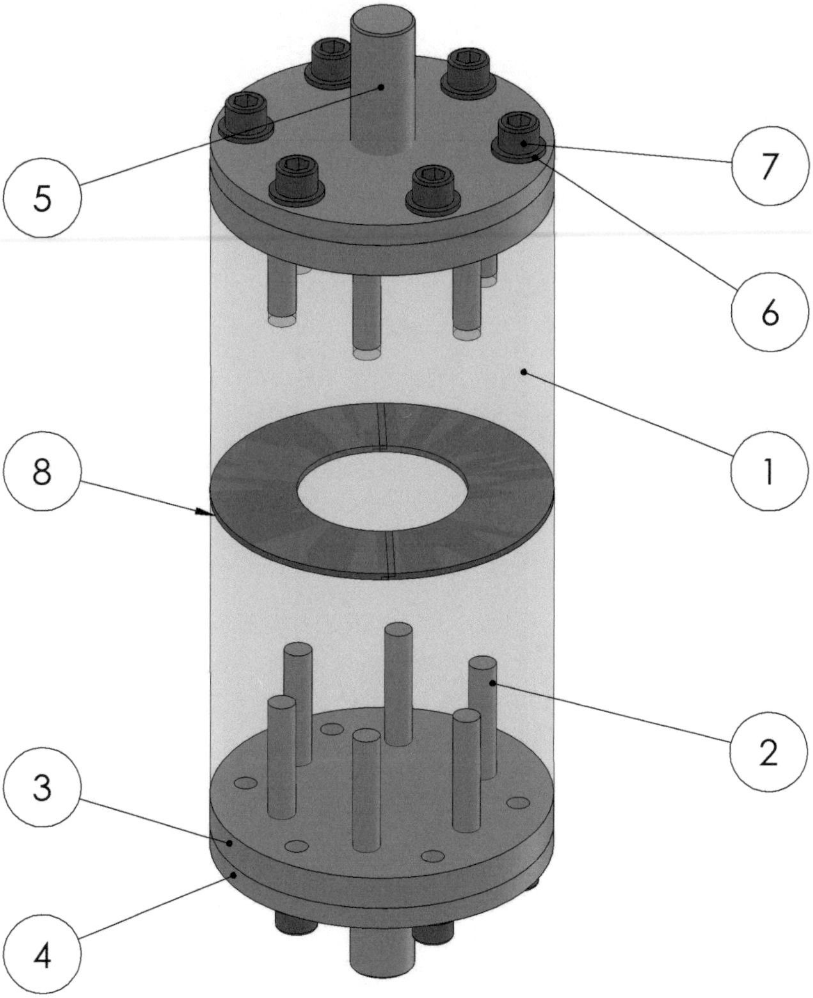

Figure 9. The device for fracture toughness test in the mode III fracture: 1—concrete specimen, 2—bolt M12/65, 3—bottom plate, 4—top plate, 5—bolt M28/70, 6—washers, 7—bolt M12/20, 8—insert.

minutes. On the basis of the experimental results, fracture toughness K_{IIIc} (**Table 11**) can be calculated, which was determined according to [38].

5.4. Determination of generalized fracture toughness

With the known values of K_{Ic}, K_{IIc}, and K_{IIIc}, it was possible to designate a generalized fracture toughness of the material K_c from the equation (1). On the basis of average values of critical stress intensity factors: K_{Ic}^S, K_{IIc} and K_{IIIc} in **Table 11** and **Figure 11** show results of generalized fracture toughness, for each of the analyzed concretes, in all time periods. In addition, **Table 11** summarizes the parameters of fracture mechanics obtained in tests for all three mode fractures.

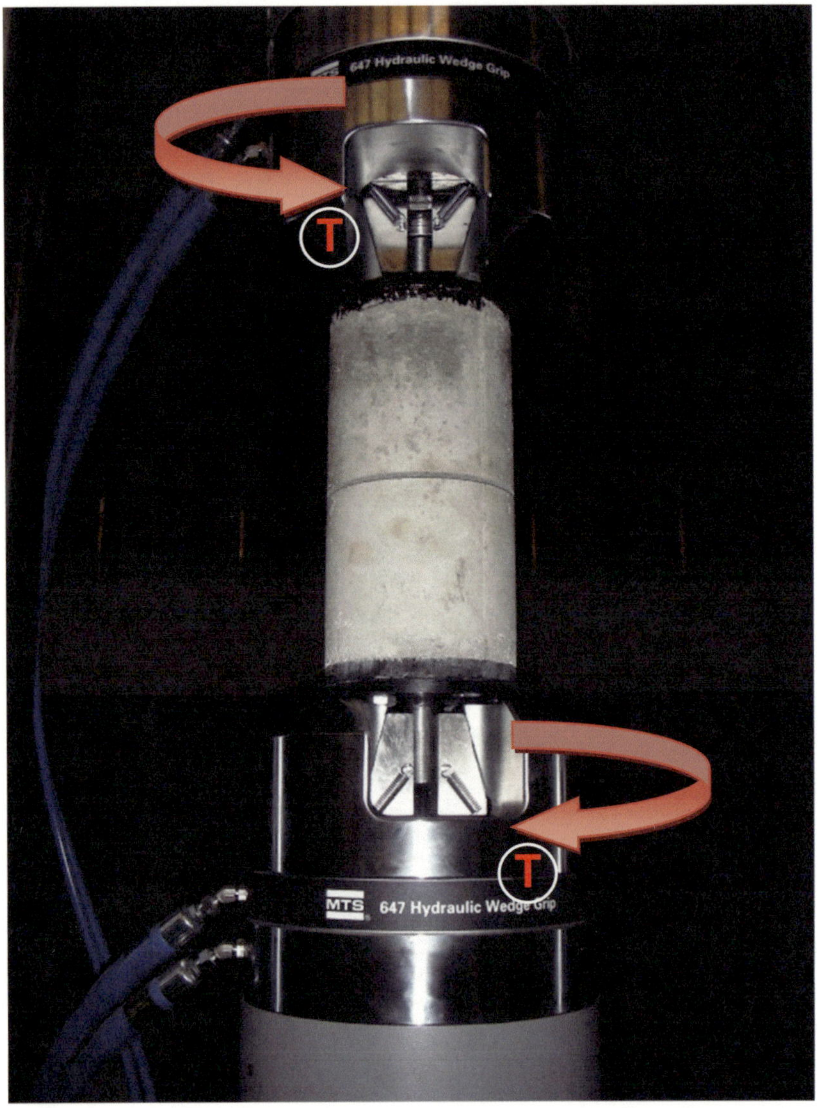

Figure 10. The experimental stand with the fixed specimen subjected to torsion; T, torque.

Based on the obtained results (**Table 11** and **Figure 11**), it can be concluded that after 3 and 7 days, the highest fracture toughness had concrete FA-00, while in the other four periods of time—FA-20. Concrete FA-30 had the lowest values of K_c in the range from 3 to 90 days, and after 180 and 365 days, the fracture toughness for this concrete is second to the results obtained for the composite with 20% siliceous fly ash.

The dynamics of changes of generalized fracture toughness in time (**Figure 12**) shows a very rapid increase of K_c after 3 and 7 days in the reference concrete and the delay of the processes

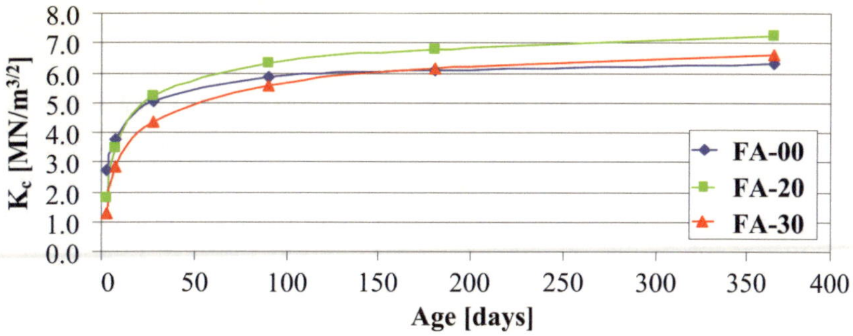

Figure 11. Generalized fracture toughness of analyzed concretes as a function of curing time.

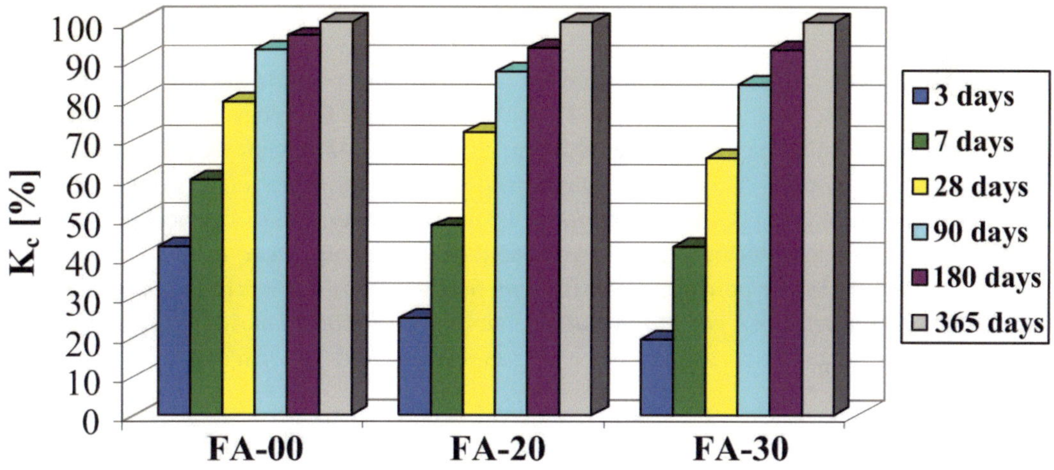

Figure 12. Relative changes of generalized fracture toughness over time.

for concretes with additives. Concrete FA-00 after 3 days reached more than 40% of the final value, while in the concretes with siliceous fly ash, it was less than 25% for FA-20 and less than 20% for FA-30. Also, after a week, large disproportions were noticed in the obtained results. The value of K_c in concrete FA-00 accounted for almost 60% of the annual result. At the same time, in concretes with siliceous fly ash, generalized fracture toughness was less than 50% K_c obtained for these concretes after 365 days. In the analysis of changes in values K_c in subsequent periods in which the tests were conducted, a steady increase of this parameter in comparative concrete as well as rapid increase in modified concretes is observed. As a result of the occurring processes, generalized fracture toughness exceeded 90% of the annual values after 6 months of curing in all of the analyzed concretes.

6. Discussion

Based on the results shown in **Table 11**, it can be concluded that siliceous fly ash additive clearly changes its generalized fracture toughness. At early age, the structure of materials with these components significantly weakens. With the 20% fly ash additive, parameter K_c after 3 days was 33.7% lower while after 7 days, it was 7.67%, in relation to values obtained for FA-00. More evident change of generalized fracture toughness resulted in an increase of the amount of the cement substitute in concrete by 10%. This resulted in a further decrease of K_c by almost 20% in 3- and 7-day concretes in comparison to results for comparative concrete. Another analysis of K_c already done for mature concrete also reflects the negative impact of 30% fly ash additive, but in relation to FA-20/28, it shows a completely different trend. Twenty percent siliceous fly ash additive after a month of curing causes an increase in fracture toughness by 3.78%, while 30% additive causes a clear drop of K_c by 13.92%. This shows that after 28 days, the 20% fly ash additive causes beneficial effects resulting from the pozzolanic reaction, whereas above this value, there is still a decrease of mechanical parameters in modified concretes.

The analysis of generalized fracture toughness in subsequent periods of time indicates that for 90-day concretes, another increase of K_c in concretes with fly ashes can be observed to the extent that the value of this parameter obtained for FA-20 is greater by 8% in relation to the value obtained for FA-00. When comparing the value of K_c for the reference concrete and FA-30, it is concluded that K_c is still greater in FA-00. However, the difference is slight and is only 5%. For concretes tested after 180 and 365 days, beneficial effect of fly ash additive is in both smaller and larger percentages of its content. Generalized fracture toughness in concretes with ashes already exceeds the values obtained for FA-00/180 and FA-00/365 by a few and more than 10% for FA-20/180 and FA-20/365 as well as 1 and 5% with respect to FA-30/180 and FA-30/365.

When comparing the effect of siliceous fly ashes on values of generalized fracture toughness (**Table 11**) and compressive strength of concretes (**Table 10**), distinct similarities in the effect of used mineral additives on the results obtained in both tests can be observed.

7. Conclusions

On the basis of comprehensive fracture toughness tests, in which a portion of the binder was replaced with active pozzolanic siliceous fly ashes, it can be concluded that:

- the siliceous fly ash additive in the amount of 20 and 30% of mass of cement significantly affects the change of fracture toughness in tension, shear, and torsion,
- obtained parameter values of fracture mechanics depend on the age of concrete,
- siliceous fly ash additive in the amount of up to 30% of mass of cement drastically reduced the fracture toughness at early age,
- 20% siliceous fly ash additive ensures high fracture toughness in mature concretes,

- concretes with 30% siliceous fly ash additive are characterized by highest dynamic increase of the parameter K_c,
- after 180 and 365 days, fracture toughness of FA-30 concrete is higher in comparison to the values obtained for FA-00 concrete.

Author details

Grzegorz Ludwik Golewski

Address all correspondence to: g.golewski@pollub.pl

Lublin University of Technology, Lublin, Poland

References

[1] Mehta PK. Greening of the concrete industry for sustainable development. Concrete International. 2002;(July):23-28

[2] Mehta PK. Reducing the environmental impact of concrete. Concrete International. 2001;(October):61-66

[3] Keun-Hyeok Y. Assessment of CO_2 reduction of alkali-activated concrete. Journal of Cleaner Production. 2013;**39**:265-272. DOI: 10.1016/j.jclepro.2012.08.001

[4] Jaya prithika A, Sekar SK. Mechanical and fracture characteristics of Eco-friendly concrete produced using coconut shell, ground granulated blast furnace slag and manufactured sand. Construction and Building Materials. 2016;**103**:1-7. DOI: 10.1016/j.conbuildmat.2015.11.035

[5] Aitcin PC. Cements of yesterday and today: Concrete of tomorrow. Cement and Concrete Research. 2000;**30**(9):1349-1359. DOI: 10.1016/S0008-8846(00)00365-3

[6] Aksogan O, Binici H, Ortlek E. Durability of concrete made by partial replacement of fine aggregate by colemanite and barite and cement by ashes of corn stalk, wheat straw and sunflower stalk ashes. Construction and Building Materials. 2016;**106**(1):253-263. DOI: 10.1016/j.conbuildmat.2015.12.102

[7] Ahmaruzzaman M. A review on the utilization of fly ash. Progress in Energy and Combustion Science. 2010;**36**(3):327-363. DOI: 10.1016/j.pecs.2009.11.03

[8] Golewski GL, Sadowski T. An analysis of shear fracture toughness K_{IIc} and microstructure in concretes containing fly-ash. Construction and Building Materials. 2014;**51**:207-214. DOI: 10.1016/j.conbuildmat.2013.10.044

[9] Blissett RS, Rowson NA. A review of the multi-component utilisation of coal fly ash. Fuel. 2012;**97**:1-23. DOI: 10.1016/j.fuel.2012.03.024

[10] Davis RE, Carlson RW, Kelly JW, Davis HE. Properties of cements and concretes containing fly ash. ACI Journal. 1937;**33**(5):577-612

[11] Berry EE, Malhotra VM. Fly ash for use in concrete—A critical review. ACI Journal. 1980;**77**(2):59-73

[12] Joshi RC, Lohtia RP. Fly Ash in Concrete. Production, Properties and Uses (Advances in Concrete Technology. Vol. 2). Amsterdam: Gordon and Breach Science Publishers; 1997. p. 271

[13] Sear LKA. Properties and Use of Coal Fly Ash. A Valuable Industrial by-Product. London: Thomas Telford Ltd.; 2001. p. 261

[14] Wesche K. Fly Ash in Concrete. Properties and Performance. London: E & FN Spon; 1991. p. 283

[15] Sadowski T, Golewski G. Effect of aggregate kind and graining on modelling of plain concrete under compression. Computational Materials Science. 2008;**43**(1):119-126. DOI: 10.1016/j.commatsci.2007.07.037

[16] Golewski G, Sadowski T. Fracture toughness at shear (mode II) of concretes made of natural and broken aggregates. In: Brandt AM, Li VC, Marshall IH, editors. Brittle Matrix Composites 8; 23th–25th October; Warsaw. Cambridge: Woodhead Publishing Ltd; 2006. pp. 537-546

[17] Santosh M, Ghosh MA. Multi-scale identification of concrete material parameters. Theoretical and Applied Fracture Mechanics. 2015;**75**:8-15. DOI: 10.1016/j.tafmec.2014.09.005

[18] Ayatollahi MR, Saboori B. T-stress effects in mixed mode I/II/III brittle fracture. Engineering Fracture Mechanics. 2015;**144**:32-45. DOI: 10.1016/j.engfracmech.2015.06.070

[19] Song L, Huang SM, Yang SC. Experimental investigation on criterion of three-dimensional mixed-mode fracture for concrete. Cement and Concrete Research. 2004;**34**(6):913-916. DOI: 10.1016/j.cemconres.2003.10.013

[20] Knott JF. Fundamentals of Fracture Mechanics. London: Butterworths; 1976. p. 273

[21] Lam L, Wong YL, Poon CS. Effect of fly ash and silica fume on compressive and fracture behaviours of concrete. Cement and Concrete Research. 1998;**28**(2):271-283. DOI: 10.1016/S0008-8846(97)00269-X

[22] Tang WC, Lo TY, Chan WK. Fracture properties of normal and lightweight high-strength concrete. Magazine of Concrete Research. 2008;**60**(4):237-244. DOI: 10.1680/macr.2008.60.4.237

[23] Bharatkumar BH, Raghuprasad BK, Ramachandramurthy DS, Narayanan R, Gopalakrishnan S. Effect of fly ash and slag on the fracture characteristics of high performance concrete. Materials and Structures. 2005;**38**(275):63-72. DOI: 10.1007/BF02480576

[24] Ganesh Babu K, Siva Nageswara Rao G. Efficiency of fly ash in concrete with age. Cement and Concrete Research. 1996;**26**(3):465-474. DOI: 10.1016/S0008-8846(96)85034-4

[25] Ganesh Babu K, Siva Nageswara Rao G. Early strength behaviour of fly ash concretes. Cement and Concrete Research. 1994;**24**(2):277-284. DOI: 10.1016/0008-8846(94)90053-1

[26] Golewski GL, Sadowski T. Experimental investigation and numerical modelling fracture processes under Mode II in concrete composites containing fly-ash additive at early age. Solid State Phenomena. 2012;**188**:158-163. DOI: 10.4028/www.scientific.net/SSP.188.158

[27] Golewski GL, Golewski P, Sadowswki T. Numerical modelling crack propagation under Mode II fracture in plain concretes containing siliceous fly-ash additive using XFEM method. Computational Materials Science. 2012;**62**:75-78. DOI: 10.1016/j.commatsci.2012.05.009

[28] Golewski GL, Sadowski T. Macroscopic evaluation of fracture processes in fly ash concrete. Solid State Phenomena. 2016;**254**:188-193. DOI: 10.4028/www.scientific.net/SSP.254.188

[29] Golewski GL, Sadowski T. A study of mode III fracture toughness in young and mature concrete with fly ash additive. Solid State Phenomena. 2016;**254**:120-125. DOI: 10.4028/www.scientific.net/SSP.254.120

[30] Chindaprasirt P, Jaturapitakkul C, Sinsiri T. Effect of fly ash fineness on compressive strength and pore size of blended cement paste. Cement and Concrete Composites. 2005;**27**(4):425-428. DOI: 10.1016/j.cemconcomp.2004.07.003

[31] Lee HJ, Lee JH, Kim DG. Study on the change in microstructure of fly ash concrete depending on ages and degree of hydration using XRD and SEM. Advanced Materials Research. 2012;**486**:350-355. DOI: 10.4028/www.scientific.net/AMR.486.350

[32] Ogawa K, Uchikawa H, Takemoto K, Yasui I. The mechanism of the hydration in the system C3S-pozzolana. Cement and Concrete Research. 1980;**10**(5):683-696. DOI: 10.1016/0008-8846(80)90032-0

[33] Fraay ALA, Bijen JM, de Haan YM. The reaction of fly ash in concrete a critical examination. Cement and Concrete Research. 1989;**19**(2):235-246. DOI: 10.1016/0008-8846(89)90088-4

[34] Zhang MH. Microstructure, crack propagation, and mechanical properties of cement pastes containing high volumes of fly ashes. Cement and Concrete Research. 1995;**25**(6):1165-1178. DOI: 10.1016/0008-8846(95)00109-P

[35] Golewski GL. Studies of natural radiactivity of concrete with siliceous fly ash addition. Cement Lime Concrete. 2015;**20**(2):106-114

[36] Shah SP, RILEM TC 89-FMT. Determination of fracture parameters (KSIc and CTODc) of plain concrete using three-point bend tests. Materials and Structures. 1990;**23**(138):457-460. DOI: 10.1007/BF02472029

[37] Watkins J. Fracture toughness test for soil-cement samples in mode II. International Journal of Fracture. 1983;**23**(4):R135-R138

[38] Miannay DP. Fracture Mechanics. New York: Springer-Verlag; 1998. p. 319